Alfred J Pearce

The science of the stars

Alfred J Pearce

The science of the stars

ISBN/EAN: 9783742814197

Manufactured in Europe, USA, Canada, Australia, Japa

Cover: Foto ©ninafisch / pixelio.de

Manufactured and distributed by brebook publishing software
(www.brebook.com)

Alfred J Pearce

The science of the stars

THE SCIENCE OF THE STARS.

BY

ALFRED J. PEARCE.

———

Καί φλογωπὰ σήματα
ἐξωρμάτωσα, πρόσθεν ὄντ' ἐπάργεμα.

ÆSCHYLUS.

" And I brought to light the fiery symbols that were aforetime wrapt in darkness."

———

LONDON:

SIMPKIN, MARSHALL, and CO.,

4, STATIONERS' HALL COURT.

———

1881.

THE most ancient of all sciences—the Science of the Stars—is but very imperfectly understood at the present day.

Modern astronomers, while improving astronomy, reject the ancient astrology. They accept the Pythagorean (under the title of the Copernican) system of astronomy, they recognise the truth and importance of Kepler's laws, yet they reject as unworthy even of examination the belief in planetary influence to which Pythagoras and Kepler subscribed. The very fact that men of such giant intellects as those two great philosophers, found, as the result of their experience and observation, that planetary influence is exerted on the atmosphere, etc., of the earth, should command respect for astrology and should lead to an impartial and thorough examination of it.

The highly educated portion of the public too often sneer at judicial astrology as mere superstition, and excuse themselves from any investigation of its claims to be considered a science, on the ground that modern astronomers reject it. They look upon it as a relic of the bizarre superstition of the ancients, and of the mediæval alchemists. If authority alone could be allowed to decide the vexed question of the truth of astrology, it will be recognised, from the names of the founders of modern astronomy before quoted, that the weight of authority is decidedly in favour of astrology. But the truth of no science can be allowed to be decided by authority alone. Hence the author offers the intelligent public the following epitome of astrology, which is written in a perfectly clear manner, free from abstruse terms, and free from the superstitious nonsense too often to be found in astrological literature. This work is based on the results of an impartial and most searching examination of the subject, extending over twenty-two years.

The meteorological observations of the author, and those of some friends of his who have investigated the subject, lead to the conclusion that no real progress will ever be made in forecasting weather until planetary influence shall be recognised.

In like manner, experience shows that an examination of astrology throws a flood of light on mental gifts, on the subject of insanity, and, to a certain extent, on "the ills that flesh is heir to."

This work is, then, offered to the world with the sole desire to spread a knowledge of divine truth, and to open up a view of one of the harmonies of the universe.

London: November 23, 1881.

CHAPTER I.

INTRODUCTION.

"In natural science there is one language universally intelligible, the language of facts; it belongs to nature, and it is as permanent as the objects of nature."—SIR HUMPHRY DAVY.

THE SCIENCE OF THE STARS is at once the most exalted and the most fascinating of all sciences. It was formerly designated ASTROLOGY—from αστηρ (aster), a star, and λογος (logos), reason, logic, or information—and it comprised the foretelling of the return of the planets, the eclipses of the Sun and Moon, tempests, droughts, inundations, earthquakes, the rise and fall of nations, wars, revolutions, the destinies of remarkable men, etc. It was this science which Dante declared to be "the highest and noblest, and without defect." It was this science in which Pythagoras, Anaxagoras, Anaximander, Democritus, Thales, Eudoxus, Hippocrates, Galen, Nigidius Figulus, Kepler, Lord Bacon, Dryden, and many other great men, were skilled. The modern system of ASTRONOMY—from αστηρ and νομος (nomos) a law—is not so comprehensive, being limited to the study and demonstration of the laws that govern the motions of the heavenly bodies. Among the ancient Hebrews the astrologer was called ASH-PHE, literally "the mouthpiece of the star," because he interpreted what he conceived to be the import of the configurations of the stars—κατα λογον, in conformity with reason.

B

It is the common cant of the day to say that the Copernican system of astronomy overthrew the ancient system of astrology, and even Mr. R. A. Proctor repeats this assertion. A little reflection will show that this is an utterly mistaken idea. Pythagoras—whose giant intellect has, perhaps, never been equalled,—who anticipated the discoveries of Copernicus, accepted the belief in astrology prevailing in his day. Kepler—"the legislator of the heavens"—avowed his belief in astrology, and constantly practised it. Lord Bacon accepted it, under the designation of *Astrologia Sana*, as a part of physics. Mr. Proctor endeavours to discount the value of Bacon's authority in its favour on the plea that he was a Ptolemaist. "Ptolemy's order, false as it was, enabled observers to give a plausible account of the motions of the Sun and Moon, to foretell eclipses, and to improve geography."[1] Inasmuch as the Ptolemaic system of astronomy pourtrayed the actual phenomena of the heavens as they appear to observers on the earth, it follows that his astrology is quite as applicable to modern and improved astronomy as to his own ; for the heavenly bodies act upon the earth, its atmosphere, and (directly or indirectly) on mankind, according to their apparent or *geocentric* positions. The great distances of the planets—as assigned to them in the Newtonian system—must not be allowed to prejudice our minds against the belief in "planetary influence." Astronomers teach that each planet is always attracting its fellows away from the average path round the Sun.

It is true that as astronomy was improved, and as astronomers received State appointments, astrology declined. This coincidence was not due to any want of

---

[1] *Spectacle de la Nature.*

truth in an *astrologia sana ;* it was the result of the
corruption of astrology by the mediæval astrologers who
degraded it into a system of mere fortune-telling; and
the astonishing advance of other sciences opening up
lucrative pursuits for clever men, astrology was abandoned to the illiterate. Nevertheless, it is on record
that the first astronomer-royal, Flamstead, practised and
believed in astrology, for he selected an auspicious
moment for the laying of the foundation-stone of Greenwich Observatory (when Jupiter, " the greater fortune,"
was rising), and a map of the heavens drawn by his own
hand may be seen among his MSS., carefully pre-
served at the Royal Observatory.[2] Some learned
astronomer, in evident ignorance of its meaning, has
pencilled in the words " *risum teneatis amici.*"

It is also on record that Newton became attracted to
the study of mathematics and astronomy by the contem-
plation of an astrological figure of the heavens.

The astronomers of the present day appear to know
little or nothing of astrology. The late Sir David
Brewster said that—" In attempting to reduce astrology
to the form of a science, there can be little doubt that
the inductive method was never followed." Mr. Proctor
has expressed his impression that the ancient astrologers
" guessed the influences of special planets from colour,
appearance, motion, etc., and having no real knowledge
to check them, they formulated what they supposed to
be a system." These utterly mistaken notions are completely refuted by the disclosures of the cuneiform inscriptions, which prove that the phenomena of the
weather were observed and recorded together with the

<hr>

[2] A *fac simile* of this map is given in " The Text-Book of
Astrology," vol i., p. 20.

B 2

configurations of the heavenly bodies. M. Lenormant says of the Chaldæans: "On nota les coïncidences qui se produisaient entre les positions ou les apparences des astres et les événements et l'on crut trouver dans ces coïncidences la clef des prévisions de l'avenir. Dés lors, l'astrologie était fondée." It is patent to every careful student of the subject that astrology was based upon a long series of careful observations. Much of the prejudice against astrology arises from the prevalent misconception as to the meaning of the word "aspect" as applied to the stars. Here is Kepler's definition: "*Aspectus est angulus à radiis luminosis binorum planetarum in terra formatus, efficax ad stimulandum naturam sublunarem*"—"An aspect is an angle formed on the earth by the luminous beams of two planets, of strength to stir up the virtue of sublunary things." Thus when the planet Jupiter is 90° distant in geocentric longitude from the Sun, it is said to be in square *aspect* with the Sun, as on the 17th of August, 1881—marked in the *Nautical Almanac* as ♃ □ ☉.

Now that observation has shown the sun-spot period, the true magnetic declination period, and that of the auroral displays to be 11.9 years, which is the same as Jupiter's anomalistic year or the time that elapses between two perihelion passages[3]; now that a connexion is demonstrable between the fluctuations of the annual death-rate and the position of Jupiter in his orbit[4]; now that the *maxima* and *minima* of earthquakes are found to synchronise with certain relative

---

[3] *Nature*, Jan. 31, 1878, letter from Mr. B. G. Jenkins, F.R.A.S.

[4] Statistical Society's *Journal*, March 1879, article by Mr. Jenkins.

positions of Jupiter and Saturn, as shown by M. Delauney, of the French Academy—some astronomers begin to recognise the fact that there was a substratum of truth in the despised system of astrology. The approach of four large planets to their perihelia in the period 1880–1885, a coincidence that has not happened for five hundred years, has so over-excited the imaginations of some astronomers, ignorant of astrology, that the most extravagant and alarming predictions have been made by them as to the probable effects on the earth, its atmosphere, and its inhabitants. Any effects traceable to the perihelion passages of the larger planets must be due to their being nearer than when in any other parts of their orbits, and would go to prove that such planets are always acting magnetically and electrically, more or less.

Observers note repeated coincidences between certain "aspects" of the heavenly bodies and great events: and, by an empirical law, we may foretell that when similar configurations shall recur, similar events will coincide or immediately follow; but in the present state of our knowledge, we cannot explain the *modus operandi* —"*causa latet, res est notissima,*" the cause is hidden, the effect most plain.

Faraday said: "The philosopher should be a man willing to listen to every suggestion, but determined to judge for himself. He should not be biassed by appearances, have no favourite hypothesis, be of no school, and in doctrine have no master. He should not be a respecter of persons but of things. Truth should be his primary object. If to these qualities be added industry, he may indeed hope to walk within the veil of the temple of Nature."

While prejudice is allowed to bar the way of scientific inquiry into the vexed question of planetary influence, the modern philosopher will never " walk within the veil of the temple of Nature." Those who have cast aside their prejudices, and have investigated the subject, have found much truth in it, and great encouragement to prosecute their inquiry with the utmost diligence. To one who is utterly unacquainted with astrology, and yet speaks or writes concerning it in an abusive and disrespectful manner, we commend the rebuke administered by Newton to Halley, " I have studied these things ; you have not."

No religious scruples should deter the student from at least the perusal of this little volume. *Astrologia sana* has nothing whatever in common with palmistry, card-shuffling, spirit-rapping, or witchcraft. It does not lead towards atheism, fatalism, &c. Lord Bacon said :— " There is no fatal necessity in the stars, and this the more prudent astrologers have allowed." Placidus, the Italian monk, author of the system of genethliacal astrology which bears his name, declared that he was " wholly of the opinion that every man is the author of his own fortune, next, however, to the Divine decree, according to the prophet, 'My lot is in thine hand.'" The Bible[5] is replete with astrology : Abraham, Job, Jacob, Moses, Joseph, David, Solomon, Daniel, and Ezekiel practised it ; Joseph even practised divination (*horary* astrology), a branch rejected by Bacon and several eminent astrologers. A Cardinal of the Catholic Church was the author of a treatise on astrology containing an *Ephemeris* for several years, a copy of which may be seen at the British Museum. Melancthon upheld astrology. It is said that Luther condemned it— perhaps owing to the very evil horoscope assigned to

---

[5] See "Veritas," by H. Melville ; and the *Anacalypsis*, by Godfrey Higgins.

him by the great Cardan. Phrenologists will understand that Melancthon's judgment on a scientific subject is entitled to far greater weight and respect than Luther's.

All ancient religions had a common basis in astrology. The ancients believed that the planets (or planetary angels) had under their special care the affairs of mankind. Phornutus (Περι Ουρανος) says: "For the ancients took those for gods whom they found to move in a certain and regular manner, thinking them to be the causes of the changes of the air and the conservation of the universe. These then are gods (δεοι), which are the disposers (δετηρες) and formers of all things."

The first verse of Genesis reads, literally rendered, thus: "In the first place God put together the original matter of the disposers [or planets], and the original matter of the earth."

The adoption by Moses of the astrological emblems of the Magi and the Egyptians clearly proves the identity of his religion with that of the Magi before it became corrupted.

The 19th Psalm refers to the "rule," or more correctly the "line," of the planets which "has gone out through all the earth."

In Matthew xxiv., 29, and in Mark xiii., 25, the Saviour alluded to the "powers" (or virtues) in the heavenly bodies, or stars, &c.

With these introductory remarks we will now proceed to delineate the various branches of judicial astrology, leaving it to the good judgment of the intelligent reader whether it deserves further and more extended examination and study.

CHAPTER II.

MUNDANE ASTROLOGY.

> " Speculataque longe
> Deprendit tacitis dominantia legibus astra,
> Et totum alterna mundum ratione moveri,
> Fatorumque vices certis discernere signis."
>
> MANILIUS.

MUNDANE ASTROLOGY relates to the forecasting of the great events and changes of the world—wars, revolutions, etc.

To this end several methods are employed :—1. By casting " figures of the heavens " for the moment of the SUN's entry into the cardinal signs (*Aries, Cancer, Libra,* and *Capricornus*). 2. By casting figures of the heavens for ECLIPSES of the Sun and Moon—in countries where they are visible. 3. By marking the transits of the superior planets (Mars, Jupiter, Saturn, Uranus, and Neptune) through the various "signs of the zodiac." 4. By watching the progress of certain eminent fixed stars through the signs of the zodiac. 5. By the observation of comets. Let us examine these several methods in detail.

I. The entry of the SUN into *Aries* is the commencement of the astrological year (March 20th). It is remarkable that the Christian era is connected with the epoch of the vernal equinox in *Aries* (the sacrificial ram, or *lamb*). At the British Museum there is a colossal figure of a Bull, with the face of a man, with wings ; similar figures are found in India, Assyria, and Egypt, and sculptured on the ancient temples. It represents the Sun at the opening of the year (vernal equinox). This carries us back more than six thousand years; it is more than four thousand years since the vernal equinox left *Taurus* and entered *Aries*. During the 2,160 years that the vernal equinox fell in the sign *Gemini*, the Sun was named and worshipped Buddha; when it fell in *Aries*, the Sun was worshipped as *Krishna*. The Cherubim mentioned in the Book of Ezekiel are simply the faces of the four Beings who were at the four cardinal points, beginning with the Sun in *Taurus*, *viz.* the ox, the lion, the eagle, and the man[1] (♉, ♌, ♏ and ♒).

Claudius Ptolemy, who wrote the *Tetrabiblos*, or Four Books of the Stars, said : " The beginning of the whole zodiacal circle (which in its nature as a circle can have no other beginning or end, capable of being determined) is, therefore, assumed to be the sign of *Aries*, which commences at the vernal equinox." Thus, the " precession of the equinoxes" cannot affect astrology. Ptolemy considered the virtues of the *constellations* of the zodiac distinctly from those of the *spaces* they occupied. He expressly and repeatedly declared that the point of the vernal equinox is the beginning of the zodiac, and that the thirty degrees immediately following it ever retain the same virtue as that which he attributed to *Aries*, although the stars forming *Aries* may have left those degrees. It was an accepted belief for ages that the world was created at the vernal equinox. Hence the astrological belief arose that the vernal equinox was the " revolution" of the world.

A map of the heavens is, then, drawn for the moment of the Sun's ingress into *Aries* (for the seat of government). If a " fixed " sign (*Taurus, Leo, Scorpio,* or

---

[1] See a picture of them in Parkhurst's Heb. Lex.

Aquarius) ascend thereat, the configurations are believe to remain in force for the ensuing twelve months. If a "common" sign (*Gemini, Virgo, Sagittarius,* or *Pisces*) ascend, the figure is held to rule for six months, and another must be drawn for the entry of the Sun into *Libra* (autumnal equinox). If a "moveable" sign (*Aries, Cancer, Libra,* or *Capricornus*) ascend, the figure is held to retain influence but for three months, and a map must be drawn for the summer solstice, autumnal equinox, and winter solstice. Modern astrologers do not observe these instructions; they cast a figure for each of the four quarters of the year. The new or full moon nearest the vernal equinox is also believed to pre-signify events about to happen.

Before proceeding further, it will be well to describe the most approved method of casting a "figure of the heavens." It is usually drawn in the form of a circle, which is divided into twelve "mansions" or "houses." It has two hemispheres, the upper diurnal and the lower nocturnal. It has two other grand divisions, effected by the line drawn from the upper meridian (where the Sun is at noon) to the lower (where the Sun is at midnight). These four divisions are the east, south, west, and north *angles*, respectively. The following diagram will show at a glance the signification attached to the several "houses."

The inner circle represents the earth, the outer the heavens. The double line drawn from E. to W. is the horizon line; the line from S. to N. is drawn from the upper to the lower meridian. Each quadrant is sub-divided into three equal parts. The ascendant is the first house, and the other houses are numbered in rotation as shown in the diagram. The Sun, in his daily course, passes through the 12th, 11th, 10th, 9th, 8th, and 7th houses; in his nightly course, through the 6th,

5th, 4th, 3rd, 2nd, and 1st houses. The 1st, 4th, 7th, and 10th houses are the "*angles ;*" a planet located in any one of them, at a Solar ingress, an eclipse, a great conjunction, or a birth, is considered to be very powerful. The order of importance of the various houses

FIG. 1.

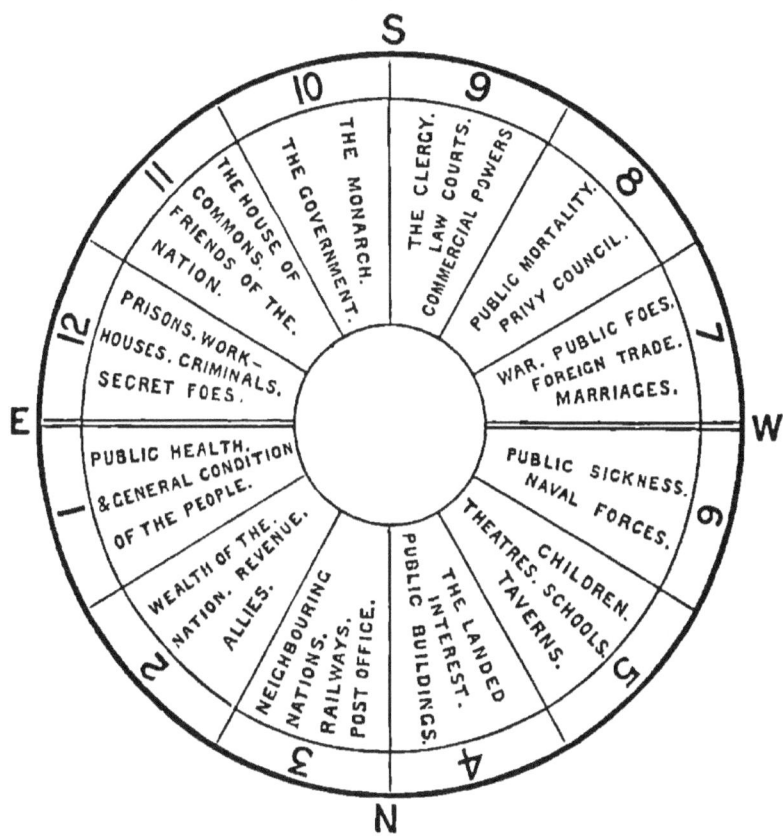

is usually stated as follows: 1st, 10th, 7th, 4th, 11th, 9th, 5th, 2nd, 3rd, 8th, 6th, and 12th. Experience shows the first house (ascendant) to be most powerful, and the upper meridian (10th) next; then follow the 7th and 4th. It is not according to reason that the

12th house should be considered the weakest of all. Next after the angles, the diurnal houses (12th, 11th, 9th, and 8th) should be considered the most powerful. The rationale of dividing the heavens is fully considered in the " Text-Book of Astrology," vol. i.

Reference to the vernal ingress of 1880 (the figure was given at p. 86 of *Urania* for March 1880), which took place at $5^h 13^m$ a.m. (G.M.T.) of March 20th, will show that the planet Mars was exactly on the lower meridian (and in opposition to the tenth house). The general election immediately following resulted in the defeat of the Conservative Government, and Lord Beaconsfield resigned in April. At Cabul Mars was rising at that ingress; fighting in Afghanistan was renewed.

At the winter solstice of 1879 (December 22nd, $4^h 16^m$ a.m., London), Mars was setting (in the sign *Taurus*). The first symptoms of that reign of terror in Ireland which unfortunately obtained in 1880 and 1881, appeared in the first quarter of 1880.

Many similar striking coincidences could be cited. While recognising the fact that the ascendant relates to the public health and the condition of the people; the tenth house to the Monarch or Government; the seventh house to war (to a certain extent) and public enemies; and the fourth to the landed interest, and the enemies of the Government in power; the signification attached to the remaining houses requires confirmation, and may be fairly questioned.

The general character of the influences ascribed to the planets, in this connection, may be summarised thus:—

JUPITER ($2\!\!\!/$) is the harbinger of peace, prosperity, reform, justice, mercy, joviality, honour, and health.

VENUS ($♀$) is the harbinger of mirth, love, feasting, pleasure, and prosperity.

MARS (♂) is the herald of war, strife, commotion, and bloodshed. In the ascendant and "strong" (*i.e.*, in *Aries, Scorpio*, or *Capricornus*), the red planet is held to presage victory for the nation concerned. In the seventh, powerful enemies, danger of war, and many troubles. In the tenth, heavy taxation, difficulties for the Government, and the ascendancy of martial men and measures. In the fourth house, mischief by fire, defeat of the Government, loss to farmers, and danger to miners. War generally follows when Mars is *retrograde* at an ingress or an eclipse, and afflicting the Sun or Moon.

SATURN (♄) is the significator of woe, misery, sickness, melancholy, etc. In the fourth house (except when in *Capricornus*) loss to farmers through bad crops, etc.

MERCURY (☿) is the ruler of science, literature, art, travelling, merchandise, etc. The influence of this planet is convertible, being good when configurated with Jupiter or Venus, and evil when with Mars, Saturn, or Uranus.

URANUS (♅) is the significator of strange, sudden, and extraordinary events; when configurated evilly with Saturn or Mars, of explosions and strange accidents; when evilly configurated with Venus or the Moon, of much evil to women and young girls.

NEPTUNE (♆) is but little understood. This distant planet is believed to be the significator of great crises—such as usually result in benefits to the countries affected.

The SUN (☉) is held to be the general significator of emperors, empresses, kings, queens, princes, princesses, presidents of republics, and all who are in supreme power and authority. Hence, when the Sun is angular in the ascendant or midheaven, free from affliction by the malefics and supported by the benefics, the prosperity, honour, and renown of the monarch or ruler are pre-signified. But, when in the descendant, the 4th, 6th, 8th, or 12th house, and afflicted by Mars, Saturn, or Uranus, evil to the ruler is threatened.

The Moon (☽) being held to be the significator of the common people, her presence in the ascendant or 10th house, unafflicted, is held to be of fortunate omen for the people; whereas her presence in the 7th, 4th, 6th, 8th, or 12th house, afflicted, is believed to be the precursor of great evils for the public.

Let us now proceed to cast a figure of the heavens for the ingress of the Sun into *Capricornus* (winter solstice), 1881.   Upon calculation, it is found that the ingress takes place at 4^h p.m. of December 21st.   To find the right-ascension of the meridian we take the

	h.	m.	s.
Sidereal time at noon	18	0	46
And add the time elapsed......... .	4	0	0
Also the difference between } mean and sidereal time for 4^h }	0	0	40
R.A. of Meridian .... =	22	1	26

A "Table of Houses for London," such as that given in the "Appendix," shows that when this amount of right-ascension is in the meridian, the 3° of the sign *Cancer* is in the ascendant; and the longitudes of the various other houses.   It will now be necessary to reduce, by proportion, the geocentric longitudes of the Sun, Moon, and planets to the hour of the ingress; and when this shall be effected, and their symbols marked in the proper places, the figure will be complete.

FIG. 2.

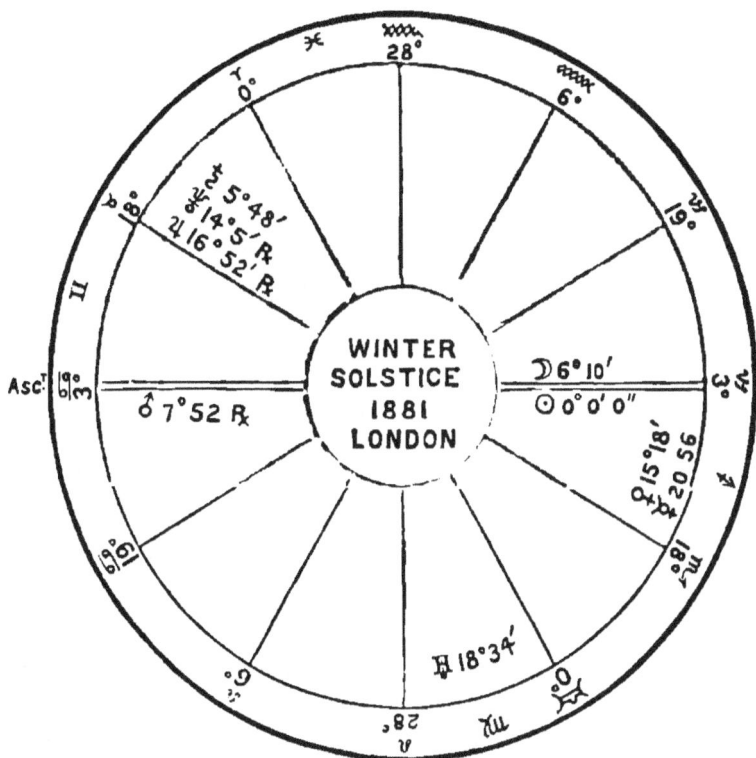

This is a very significant figure, for Mars is rising, *retrograde*, and in opposition to the Moon. Moreover, Jupiter, Saturn, and Neptune are *retrograde* in *Taurus;* Uranus is *stationary* in the sign *Virgo*, and in square (90°) aspect with Mercury and Venus. These positions pre-signify violence in the land (and especially in Africa, ruled by *Cancer*, the sign occupied by Mars), misfortunes, and accidents.

CHAPTER III.

MUNDANE ASTROLOGY—Continued.

II. The transits of the superior planets through the signs of the zodiac. The familiarity of the inhabited earth with the various signs of the zodiac was described by Ptolemy in his *Tetrabiblos*, Book II., chap. 3. No alteration has been found necessary, in the opinion of modern astrologers; and the countries that have been discovered and populated since Ptolemy's day, have (by observation) been found to have familiarity with certain signs. It is a remarkable fact that Britain and Germany are ever found to be influenced by the transits of the superior planets through *Aries;* in like manner, France and Italy are still found to be affected by *Leo*, India and Greece by *Capricornus*. Coincidences repeated through many centuries—history repeating itself—impress the mind, and lead to the conviction that there is some mysterious connection between them. No country prospers during the period of Saturn's stay in its "ruling sign;" while, on the other hand, the transits of Jupiter coincide with fortunate events.

The relationship of the several signs of the zodiac with the countries and chief cities of the world is thus stated by modern astrologers:—

Aries influences Britain, Germany, Denmark, Lesser Poland, Burgundy, Palestine, Syria, or Judea. *Towns:* Naples, Capua, Florence, Verona, Padua, Brunswick, Marseilles, Cracow, Saragossa, and Utrecht.

Taurus influences Persia, Media, Georgia, the Caucasus, Asia Minor, the Archipelago, Cyprus, Poland,

Ireland, and White Russia. Towns: Dublin, Mantua, Leipsic, Parma, Rhodes, and Palermo.

GEMINI relates to the north-east coast of Africa, to Lower Egypt, Flanders, Lombardy, Sardinia, Brabant, and Belgium; also the West of England, and the United States of North America. Towns: LONDON, Versailles, Metz, Lovaine, Bruges, Cordova, and Nuremberg; perhaps Melbourne.

CANCER rules Northern and Western Africa, Scotland, Holland, and Zealand. Towns: Amsterdam, Constantinople, Cadiz, Genoa, Venice, Algiers, Tunis, York, New York, St. Andrews, Berne, Milan, Lubeck, Vincentia, Magdeburg, and Manchester.

LEO rules France, Italy, Sicily, the Alps, Bohemia, Chaldea, the ancient Phœnicia, or the northern parts of Roumania. Towns: Rome, Bath, Bristol, Taunton, Damascus, Prague, Ravenna, Philadelphia, and probably Portsmouth.

VIRGO influences Turkey in Europe and Asia; Babylonia, Assyria; all the country between the Tigris and Euphrates; Greece, Thessaly, Corinth, and the Morea; the island of Candia, Croatia, and Switzerland. Towns: Heidelberg, Jerusalem, Paris, Reading, Lyons, Toulouse, and probably Cheltenham.

LIBRA relates to the borders of the Caspian, part of Thibet; China, especially the northern provinces; Japan, parts of India, Austria, Savoy, Upper Egypt, and ancient Libya. Towns: Antwerp, Frankfort, Lisbon, Spires, Fribourg, Vienna, Gaëta, Charlestown, and Placenza.

SCORPIO rules Fez, Morocco, Algiers, Barbary, Judea, Syria, and Cappadocia; Norway and Jutland, Bavaria, Valentia, and Catalonia. Towns: Frankfort-on-the-Oder, Liverpool, and Messina.

SAGITTARIUS influences Spain, Tuscany, lower Italy

(especially Tarento), that part of France between La Seine and La Garonne to Cape Finisterre, Arabia Felix, Dalmatia, Sclavonia, Hungary, and Moravia. Towns: Cologne, Avignon, Buda, Narbonne, Toledo, Rotenburg, and Stutgardt; probably Sunderland.

CAPRICORNUS rules India, Afghanistan, the modern Punjaub, Thrace, Macedonia, the Morea and Illyria, Bosnia, Bulgaria, Albania, Styria, Romandiola in Italy, the south-west of Saxony, Hesse, Mexico, and Mecklenburg. Towns: Oxford, Prato in Tuscany, Brandenburg, Tortona, Canstanz, and perhaps Brussels.

AQUARIUS rules Arabia the Stony, Red Russia, Prussia, part of Poland, Lithuania, Tartary, part of Muscovy, Circassia, Wallachia, Sweden, Westphalia, Piedmont, Azania, and Abyssinia. Towns: Hamburg, Bremen, Saltzburg, Trent, and Ingoldstadt.

PISCES influences Portugal, Calabria, Normandy, Galicia in Spain, Egypt, Nubia, and the southern parts of Asia Minor. Towns: Alexandria, Ratisbon, Worms, Seville, Tiverton, Farnham, and probably Bournemouth.

If we refer to the transits of Saturn through the sign *Aries* during the past six hundred years, we shall be surprised at the frequent coincidence of misfortunes to England. In 1290 Saturn was in *Aries*; the desperate war with the Scots was waged by Edward III., and the English army was defeated at Roslin near Edinburgh. In 1319, $\hbar$ in Υ, occurred the rebellion under the Earls of Lancaster and Hereford. In 1349, $\hbar$ in Υ, the Black Prince was defeated in France, and subsequently died to the great grief of the nation. In 1378, $\hbar$ in Υ, occurred the rebellion headed by Wat Tyler. In 1437, $\hbar$ in Υ, the English forces met with repeated defeats in France, losing all but Calais. In 1466, $\hbar$ in Υ, civil war was raging, headed by Warwick. In 1555, $\hbar$ in Υ, in Queen Mary's reign, 277 persons burnt at the stake. In 1584, ♄ in ♈, a plot was discovered to assassinate Queen Elizabeth; fourteen persons hanged for their participation in it. In 1643, ♄ in ♈, civil war between Charles I. and the Parliament. In 1761, ♄ in ♈, the Spaniards joined France in her war against England. In 1790, ♄ in ♈, war in India, riots at Birmingham (ruled by ♈) and such great national distress as "exceeded all that had ever happened," to use Goldsmith's words. In 1820, ♄ in ♈, trial of Queen Caroline, great national excitement, royalty disgraced, tumults, etc. In 1849, ♄ in ♈, the Asiatic cholera visited this country; there was great depression in trade; war in the Punjaub.

It is remarkable that when Jupiter was with Saturn in *Aries*, as in 1702 and in 1821, the evil influence of the latter planet was greatly mitigated.

Now let us refer to a few instances of the coincidence of Jupiter's stay in *Aries* with events favourable to England; as Jupiter is in *Aries* every twelve years, it will be impossible, for want of space, to refer to many. In 896, Jupiter was in *Aries;* King Alfred beat the Danes. In 1215, ♃ in ♈; King John was forced to sign *Magna Charta*. In 1346, ♃ in ♈; the battle of Cressy was won. In 1415, ♃ in ♈; the battle of Agincourt was gained. In 1690, ♃ in ♈, the battle of the Boyne was won. In May, 1856, ♃ in ♈, peace was signed between the Allies (England, France, Turkey, and Sardinia) and Russia, the Crimean War being brought to a glorious termination ; a splendid harvest in England was gathered in, the same year—literally " peace and plenty." In 1868, ♃ in ♈, a tide of prosperity for this country set in, and continued to flow for several years. In April, 1880, ♃ re-entered ♈, and an improvement took place in the trade and commerce of

England, which had been greatly depressed since Saturn entered *Aries*, in May 1878. But, as ♄ was still in ♈, trade was not so profitable as hoped for, and the disasters at the Cape cast a gloom over the empire.

It has been stated that *Gemini* rules the West of England, London, and the United States. It is remarkable that the rebellion of the American colonies coincided with the transit of Uranus through the sign *Gemini*; and that on the next occasion of the same planet passing through the same sign (1859 to 1866) the great American civil war raged for four years. In June 1861, a great conflagration took place in London, lasting for six weeks. In 1862, the West of England suffered fearfully from the "cotton famine," nearly a million of people being in a state of semi-starvation.

Similar coincidences can be stated in regard to other countries. For example : At the vernal ingress of 1854, Mars was *retrograde* in the sign *Virgo* ; and shortly afterwards the Czar declared war against Turkey.

The entry of Uranus into the sign *Capricornus*, in 1822, coincided with the outbreak of the revolution in Greece ; war raged in that country during the whole period of this planet's stay in the sign ♑. In 1829, Uranus left it, and when Jupiter entered the same sign (in 1830), the Allies interfered and put an end to the war.

CHAPTER IV.

MUNDANE ASTROLOGY—Continued.

THE PROGRESS OF THE FIXED STARS THROUGH THE SIGNS OF THE ZODIAC.

THAT eminent fixed star the "BULL'S NORTH HORN" (β *Tauri*), of the second magnitude, and of the nature of Mars, arrived at 17° 54' of the sign Gemini (the exact ascendant of London), in 1665-6, when the plague and fire of London took place. William Lilly, the celebrated astrologer, foretold, fifteen years beforehand, those momentous events, from that transit of the martial star. No doubt Lilly had observed the coincidence of the outbreak of plague in London in 1625 (the year of accession of Charles I.), to which no less than 35,417 persons fell victims, with the approach of the Bull's North Horn to the ascendant of the metropolis, for it had then reached 17° 20' of *Gemini*. It is worthy of remark that the civil war coincided with this approach of the martial star to the ascendant of the capital, and that in 1649, when it had arrived at 17° 40' of *Gemini*, the king was beheaded.

Nostradamus, also, foretold the fire of London, in the following lines:[1]—

"Le sang du juste à Londres fera faute,
Bruslez par feu, de vingt et trois, les six,
La dame antique cherra de place haute,
De même secte plusieurs seront occis."

---

[1] At the British Museum Library a copy may be seen of "Michel de Nostradame the elder; the true prophecies of," translated by F. de Garencières, London, 1672. Also a copy of Lilly's hieroglyphic of the plague and fire of London.

His predictions were published about the year 1555, during the reign of Queen Mary. The "ancient dame" referred to St. Paul's Church, which stood on the site of an ancient temple of Diana, built, as were all the ancient temples for the worshippers of Baal and the "heavenly host," on "a high place." St. Paul's and eighty-nine other churches, "of that same sect," were burnt.

Here is the calculation of the place of the Bull's North Horn in 1665:—

Longitude of β *Tauri*, Jan. 1st, 1879[2]   ♊ 20 53 40
Longitude of the ascendant of London   ♊ 17 54 0

_____

2 59 40

Then $2° 59' 40'' = 10780$ *seconds* of longitude, which divided by $50·25''$ (the annual motion of the fixed stars) $= 214$, the number of years since the star was in ♊ $17° 54' 0''$. From the year 1879 take 214, and the remainder is 1665.

When the martial star "ALDEBARAN" (α *Tauri*) of the first magnitude, shall arrive at $17° 54'$ *Gemini*— about 700 years hence—there will probably happen a fearful conflagration in, if not the total destruction of London. Perchance, Macaulay's New Zealander will then contemplate the ruins of the great metropolis!

That brilliant martial star REGULUS (α *Leonis*), the lion's heart, entered the sign *Leo* (in the manner described by Ptolemy) in the year 293 B.C., when the power of Rome became very fully established, more especially its religious power or that of the Pontifex Maximus. The star passed the 28th degree of *Leo* in

_____

[2] See the longitudes of eminent fixed stars given at p. 114 of the "Text-Book of Astrology," vol. i.

the year 1868, and must be considered to have then left the sign. Two years afterwards, the Italian troops entered Rome, and the Papal (temporal) power, which had only been sustained by French troops for many years, was at an end.

The sign next in order to *Leo* being *Virgo*, the star Regulus must be considered to have entered the latter sign in the year 1868. *Virgo* is the ruling sign of Paris and Turkey. In 1870, Paris suffered the horrors of the siege; and in 1871, the Communists nearly destroyed the gay metropolis of France. In 1877, the Turks suffered untold miseries from the crusade led by Russia against them. While Regulus shall remain in *Virgo* neither Turkey nor Paris can expect long-continued peace.

It is remarkable that Napoleon I—whose military genius placed France at the head of all nations, and laid Europe at her feet until the coalition of the Great Powers, aided by the brilliant generalship of Wellington, the genius of Nelson, and the invincible courage of British soldiers and sailors, eventually conquered the French—was born when the Sun was in conjunction with Regulus (in *Leo*).

CHAPTER V.

MUNDANE ASTROLOGY—Continued.

ECLIPSES OF THE SUN AND MOON.

THE belief that these phenomena—or rather the planetary positions at the moment of greatest eclipse—are a veritable "shadow of things to come," and the most important indices of future events in the countries where they are visible, is *not*, as is commonly supposed, a relic of the superstitious dread of barbarians who feared that "the dragon" was about to swallow the darkened luminary, and who used to beat drums and make horrible noises in order to frighten away the "adversary."[1]

Observers found that the countries through which the line of central eclipse passed, were subject to the calamities pre-signified by the planetary positions. Rules were thereupon formulated by the *magi*, and have been handed down to us in a very imperfect manner, for foretelling the probable effects of eclipses and naming the countries in which such effects would probably be felt. It is an insult to the memory of such men as Thales, Democritus, and others, who computed and fore-

---

[1] The word Satan, meaning adversary, is derived from *Ash*, a fire, and *Tan*, a dragon—"the fiery dragon." As the great dragon was the enemy of Apollo, the Sun, so Satan is represented as "the adversary." The legend of St. George and the Dragon is derived from this.

told eclipses, to charge them with vulgar superstition. The luni-solar period of the Chaldæans must have been based upon an immense number of very accurate observations.

The battle of Isandhlwana was fought during an annular eclipse of the Sun. The map given at p. 396 of the *Nautical Almanac* for 1879, shows the line of central and annular eclipse passing through South Africa (and Zulu-land). This eclipse was also visible (as a partial eclipse) in that very region of South America which was the scene of the Chilian War.

The following is the map of the heavens for the moment of ecliptic conjunction of the luminaries at Capetown, where a partial eclipse was visible: —

FIG. 3.

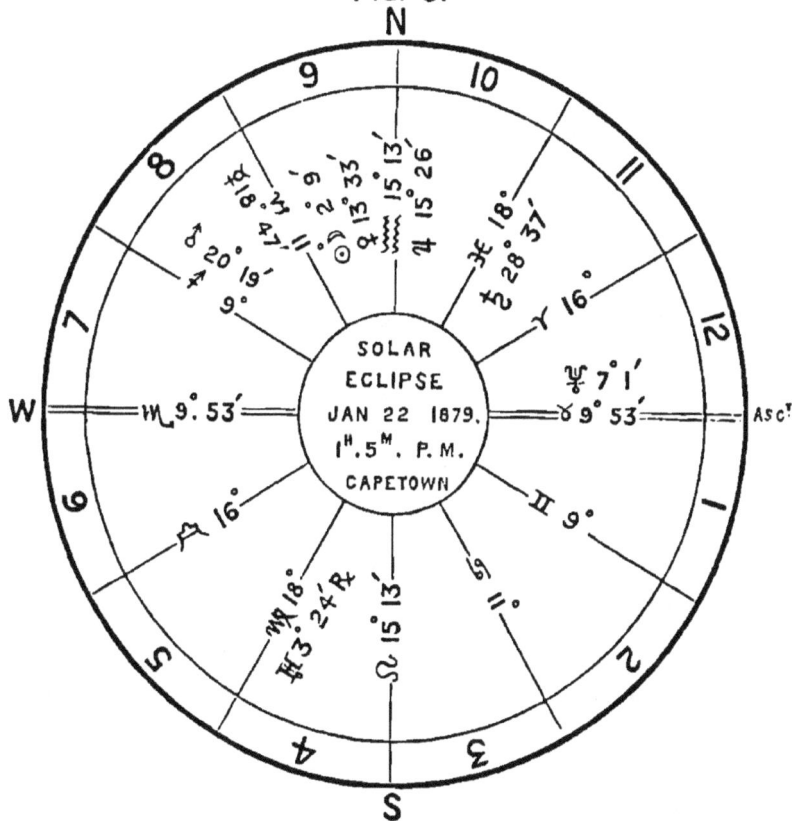

The rule for casting figures of the heavens for *southern* latitudes will be found at p. 49 of the "Text-Book of Astrology," vol. i. The lat. of Capetown is 33° 56' S., the long. 18° 28 E. of Greenwich. The pole of the 11th house is 12° 49'; and that of the 12th, 24° 22'.

The Sun was eclipsed in 2° 9' of the sign *Aquarius*. Ramesey says that " a solar eclipse falling in the first decanate of *Aquarius* causes public sorrow and sadness."

Zadkiel (*Almanac* for 1879) quoted this aphorism, and added: " Saturn in the 11th house will bring some difficulties on the Government, and these may be chiefly in connection with hostile acts perpetrated by discontented tribes, for Mars is in square (90°) aspect with Saturn."

On the 10th of April, 1865, there occurred a partial eclipse of the Moon at Washington. Jupiter was in the ascendant, in *Sagittarius*. On that very day, General Lee surrendered to General Grant, with the army of Virginia, 25,000 strong; thus putting an end to the great civil war. Zadkiel foretold this event in the following words:—" I find Jupiter strong in the ascendant, at this eclipse. I have no doubt *peace* will take place under the benefic influence of this eclipse." Unhappily, President Lincoln was assassinated four days afterwards. It is curious to observe that the Moon was in the 10th house and in conjunction with the evil Saturn, and, according to Ramesey, this pre-signified " death to some famous and illustrious man."

It may appear arbitrary to take the Moon as general significatrix (in Mundane Astrology) of the common people. Yet it would seem to have some show of reason when we remember that the Tay Bridge catastrophe, by which nearly 100 lives were lost, only one or two of the passengers ranking above the class of " common people," took place on the very evening

(December 28th, 1879) of the partial eclipse of the Moon in the sign *Cancer* (which rules Scotland)—aye, and before the shadow had entirely passed away from the Moon's disc. Ramesey avers that such an eclipse falling in *Cancer* denotes "the death and slaughter of obscure, common, plebeian kind of people."

On the 12th of July, 1870, a total eclipse of the Moon, visible in Europe, took place. Three days afterwards, Louis Napoleon declared war against Prussia. The slaughter in that war was horrible.

On the 27th of February, 1877, a total eclipse of the Moon, visible in Europe, took place in the sign *Virgo*, the Moon being in opposition to Saturn.

The relations between Russia and Turkey were then greatly strained, and on the 24th of April the late Czar declared war against the Sultan. During the height of the struggle at Plevna, another total eclipse of the Moon, visible in Europe, occurred; this time in *Pisces*, the sign opposite to *Virgo*, the Moon being attended by the two infortunes (nearly conjoined) Mars and Saturn. The holocaust of victims to the crusade against Turkey was perfectly appalling.

As to the general effects of eclipses, Ramesey asserts that—"When there happens any eclipse of the Sun or Moon in *Taurus*, *Virgo*, or *Capricornus*, it denotes a scarcity of the fruits of the earth and of corn. In *Gemini*, *Libra*, or *Aquarius*, a famine or outrageous diseases, pestilences, and mortalities. In *Cancer*, *Scorpio*, or *Pisces*, the death and slaughter of obscure, common, plebeian kind of people, continual quarrels and seditions, and great damage to navigators and sailors. In *Aries*, *Leo*, or *Sagittarius*, the motion of armies, tumults; heat, drought; troubles and anxieties to kings, princes, and magistrates; fevers, etc."

On the 25th of April, 1846, there happened an

annular eclipse of the Sun in the sixth degree of *Taurus*,[2] visible at Greenwich. This was followed by a failure of the potato crop, and great suffering in Ireland in consequence thereof. The only kind of corn which produced anything like an average return was wheat, spring corn and pulse being very deficient.

Cardan has transmitted to posterity some aphorisms relating to eclipses :—" No eclipse whatsoever can threaten a plague or scarcity to the whole earth, nor can the pestilence continue above four years in one place.

" Eclipses happening in the fourth arc stronger and more efficacious than in the eighth or twelfth house; and in the ascendant more than in the ninth or eleventh house.

" Eclipses operate more powerfully on cities, provinces, and kingdoms than on particular persons of private condition, or even upon kings and princes; for their effects rather respect the multitude.

" When eclipses happen in earthy signs (♉, ♑, ♍), they portend barrenness and scarcity, by reason of excessive droughts; when in watery signs (♋, ♏, ♓), by reason of too much rain. In airy signs (♊, ♎, ♒), they signify mighty winds, seditions, and pestilence; in fiery signs (♈, ♌, ♐), terrible wars and slaughters."

---

[2] *Vide* Zadkiel's *Almanac* for 1846, p. 35; a diagram of this eclipse is there given, and the fear expressed that there would be "a bad harvest" and "a scarcity of most fruits, especially of corn." An eclipse of the Sun in *Taurus*, visible at Dublin, took place on May 6th, 1845. There will be another on the 16th of May, 1882.

CHAPTER VI.

MUNDANE ASTROLOGY—Continued.

GREAT CONJUNCTIONS.

RAMESEY stated that " there are seven sorts of conjunctions considerable."

" 1. The first and greatest of all the rest is the conjunction of the two superior planets SATURN and JUPITER in the first term or degree of *Aries*, which happens but once in 960 years.

" 2. The second is the conjunction of SATURN and JUPITER in the first term or degree of every triplicity, and this is accomplished once in 240 years; yet once in 20 years they come into conjunction in one part or other of the zodiac.

" 3. The conjunction of SATURN and MARS in the first term or degree of *Cancer*, and this is once in 30 years.

" 4. The conjunction of the three superiors, SATURN, JUPITER, and MARS, in one term or face of any sign.

" 5. The conjunction of JUPITER and MARS, which is a mean and the least conjunction of the superiors, and therefore is not the forerunner of such great mischiefs.

" 6. The conjunction of the SUN with any planet at the time of his entrance into the first point of *Aries*.

" 7. The conjunction of the SUN and MOON, which happeneth once every month."

In consequence of the discovery of the planets URANUS and NEPTUNE, the list of great conjunctions is greatly extended, and a re-arrangement necessary.

Undoubtedly the most important conjunctions are these:—

1. The conjunction of two or more of the superior planets at or near their *perihelia*.

2. The conjunction of two or more of the superior planets when one of them is in *perigee*.

3. The conjunction of two or more of the superior planets in the first point of *Aries, Cancer, Libra*, or *Capricornus*.

The conjunction of ♂ and ♃ in 0° 1' 56" of Capricornus, on March 1st, 1877,[1] was quickly followed by the Russo-Turkish war, Bulgaria (ruled by ♑) being the scene of many sanguinary engagements. This conjunction was followed by that of Mars and Saturn in 13° 45' 21" of Pisces on the 3rd of November, 1877, when Mars was in perigee.

The greatest effects of great conjunctions generally fall on those countries and places whereat the conjoined planets are exactly rising at the moment of their conjunction, *e.g.* the conjunction of Mars and Saturn in 1879, in the ascendant at Cabul.[2]

Ramesey insists on the figure of the heavens being cast for the " punctual time of partile conjunction, *viz.*, in the very same sign, degree, and minute of the zodiac " —he should have added *seconds* also. He asserts that if the conjoined planets be strong and fortunate they presage good, if weak and impedited the contrary. This is absurd, for if an evil planet be " strong " (as Saturn is observed to be when in ♎, ♑, or ♒) it must be the

---

[1] Zadkiel foretold the war from this conjunction. Vide *Zadkiel's Almanac* for 1877. The prediction was written nine months before the declaration of war.

[2] Vide *Urania* for January, 1880.

more powerful for evil; certainly, I have never found that countries ruled by Capricornus (♑) derived any benefit from the transit of Saturn through their ruling sign. There is more sense in the following remark of Ramesey : "This good or evil shall be according to the *nature* of the planets in conjunction, and according to the nature of the sign in which they are."

It is asserted that when planets are conjoined in *fixed* signs (♉, ♌, ♍, ♒) they produce more lasting effects. I have never been able to verify this; my observations go to prove that conjunctions in the *cardinal* signs (♈ ♋, ♎, ♑) are identified with the most durable effects.

Ramesey affirms that the conjunction of Saturn and Jupiter in the first point, or first six degrees, of *Aries* is the greatest and most notable conjunction of all, and is accompanied or quickly followed by "commotions, wars, seditions, revolutions, alterations of laws, plagues, the death of kings," etc. "In like manner, the conjunction of Saturn and Mars in the first three degrees of *Cancer* is the forerunner of much evil, *viz.*, terrible wars, slaughters, depopulations, alterations of government, etc. If it be oriental, its effects will soon operate; if occidental, not so soon."

Other aphorisms, most of them arbitrary and fanciful, are given in respect to other conjunctions.

The arrangement of "terms" is altogether arbitrary, and it was made when the existence of Uranus and Neptune was unknown.

It is a fact that *Taurus* and *Scorpio* are earthquake-producing signs, *i.e.*, when containing the larger planets; *Cancer* and *Capricornus* are the same, to a less extent.

CHAPTER VII.

MUNDANE ASTROLOGY—Continued.

COMETS AS PORTENTS.

" Comets importing change of times and States,
Brandish your crystal tresses in the sky."
SHAKESPEARE.

COMETS were regarded as portents by the ancient and mediæval astrologers. The rules for judging of their signification are similar to those relating to eclipses and great conjunctions.

Albumazar and Junctinus averred that comets becoming first visible in *Aries* pre-signify evil to nobles and grandees in eastern parts and in those countries influenced by that sign; drought, drying up of fountains, etc.

The comet of 1870 appeared in *Aries* (the sign ruling Germany), and very near the degree of right ascension of the Sun and Mars at the birth of Napoleon III. Drought was experienced in most parts of Europe that year. The terrible Franco-German war was begun in July, 1870, and in the following September Louis Napoleon surrendered to the King of Prussia, at Sedan.

The appearance of Donati's comet, in 1858, was quickly followed by the Italian war of 1859. The appearance of the great comet of 1861 coincided with the great conflagration in London (which had not been equalled for two hundred years), and was immediately followed by the outbreak of the Civil War in N.

America; this comet appeared in *Gemini* (the sign ruling London and the United States).

The present year (1881) has been signalised by the appearance of three comets. The great comet (Comet *B*) was first seen, in the second decanate of *Gemini*, on the 22nd of May; on the 2nd of July, 1881, the President of the United States was shot. Great storms, perfect hurricanes, and volcanic action, quickly followed the appearance of this comet. It is singular to relate that the ancients aver that the appearance of a comet in *Gemini* pre-signifies " tempestuous, stormy winds," and " the death of some famous and illustrious man." Comet *C*, 1881, also appeared in the second decanate of *Gemini*. The great heat of the month of July, in England, France, Austria, N. America, etc., will long be remembered—being one more instance of the heat of " comet years."

The death of President Garfield—in the midheaven of whose horoscope Comet *B* appeared—the mourning for him in the United States, and the terrible forest fires, have quickly followed the appearance of these comets in *Gemini*. With these coincidences in view, we can hardly wonder at the dread felt and expressed by our forefathers at the appearance of great comets.

CHAPTER VIII.

ASTRO-METEOROLOGY.

"When the planets,
In evil mixture, to disorder wander,
What plagues! and what portents! what mutiny!
What raging of the sea! shaking of the earth!
Commotion in the winds!"—SHAKESPEARE.

THE word METEOROLOGY is derived from μετέωρος (meteoros), elevated, or soaring. The object of meteorology is stated to be "properly the scientific study of atmospheric phenomena, and the investigation of weather and climate."[1] Nothing so improper as the endeavour to trace a possible and probable connexion between *astronomical causes* and weather-changes, earthquakes, etc., is to be found in the received and State supported system of meteorology. Hence the prefix "Astro" is used to differentiate between the system presented for consideration in these pages and the received system, and to denote that astronomical causes are studied and recognised.

The disclosures of the cuneiform inscriptions prove that the phenomena of the weather were observed and recorded together with the configurations of the heavenly bodies.[2]

Kepler, Bacon, and many other eminent men of old

---

[1] Lecture by Dr. Mann, V P.M.S., delivered under the auspices of the Meteorological Society, in 1878.

[2] "Babylonian Literature." By the Rev. A. H. Sayce.

time, traced the connexion—or, to say the least, the *coincidence*—between astronomical causes and weather changes. The *savans* of the present century reject planetary influence, yet, with the exception of the improvement of meteorological instruments and the refinement of observations, they have not advanced the science.

Mr. Kendrick tells us that "by their science the Egyptian astrologers could foretell years of scarcity and plenty, pestilences, earthquakes, inundations, and the appearance of comets, and do many other things surpassing the sagacity of the vulgar."[3]

Pliny relates of Anaximander that "he foretold the earthquakes that overthrew Lacedæmon." Pliny also relates that Anaxagoras foretold the fall of a meteoric stone, about the second year of the 78th Olympiad, which occurred near the Egos, in Thrace. "It happened," says Pliny, "in sight of many, in the day-time, a comet blazing at the time, and this stone was as big as a wain could carry, and was kept for a monument."

Aristotle relates of Thales that "being upbraided by some foolish scoffers on account of his poverty, and with the unprofitableness of his studies in wisdom and philosophy, he had recourse to his astrological skill; whereby, foreseeing that in the year following olives would be unusually plentiful, to show his reproachers the vanity of their ill-timed scoffing, the winter before that year he hired all the shops and depositories (both at *Chios* and Miletum) that were reserved for the making of oil. Having got them into his hands for a very small sum, when the time came for gathering olives, every man

---

[3] "Ancient Egypt under the Pharoahs." By John Kendrick, M.A.

being destitute of rooms and offices answerable to the great plenty of olives with which they were glutted, was driven to resort to Thales for his supply thereof; who, taking advantage of that necessity, did turn them over at what price himself listed, whereby he gained a great mass of money; and afterwards, to show his contempt of riches, gave it to the poor."

What has meteorology gained by being divorced from Astrology? Can the meteorologists of the present day emulate the achievements of the Chaldæans and the ancient Egyptians—can they foretell years of plenty, pestilence, earthquakes, and inundations? No. Meteorology, so far as prediction of the weather is concerned, has retrograded rather than advanced.

In a letter to the *Times*, in the year 1875, Mr. J. G. Symons, the indefatigable hon. sec. of the B. M. Society, said—"My own shelves, and those of the library of the Meteorological Society, groan beneath the weight of masses of records; the requisite in meteorology is not observations, but brains to work out the results."

Mr. R. H. Scott, in a recent lecture, reiterated the remarks of Capt. Hoffmeyer: " When the proper time arrives, a Kepler will be surely forthcoming to discover the laws by which our science works; for us to endeavour to force the plant in its growth is hopeless."

Why, Kepler gave the key to this discovery when he averred that " A MOST UNFAILING EXPERIENCE OF THE EXCITEMENT OF SUBLUNARY NATURES BY THE CONJUNCTION AND ASPECTS OF THE PLANETS, HAS INSTRUCTED AND COMPELLED MY UNWILLING BELIEF."

Max Müller says[1]—"The torch of imagination is as

necessary to him who looks for truth, as the lamp of study; Kepler held both, and, more than that, he had the star of faith to guide him in all things from darkness to light."

The "British meteorologists" of the present day, while uttering jeremiads as to the state of meteorology, are positively too prejudiced to even examine into planetary influence. They reject with scorn, unworthy of philosophers, the experience and belief of the great Kepler; hence they are still groping in a worse than Egyptian darkness, from which they most assuredly can never emerge until they shall follow the lead of Kepler. No other guide can lead them from darkness to light. If they would but take the trouble to compare their records of observations with the "conjunctions and aspects of the planets"—*i.e.*, with the aspects formed between the Sun, Moon, Mercury, Venus, Mars, Jupiter, Uranus, and Neptune—they would quickly rescue meteorology from its present hopeless and helpless state. While they discourage such comparisons they will make no real progress whatever.

Prof. Wolf, Messrs. De la Rue, Stewart, and Loewy have done good service in directing attention to planetary action in connexion with "Solar Physics." But this is not enough; the inquiry must be pushed further, it must be made on the lines laid down by Kepler.

"Till very lately, *Caloric* was a term in constant use, and it was supposed to express some real matter, something that produced heat. That idea is now exploded, and heat is understood to be the result of *molecular* and *ethereal* vibrations."[5]

---

[5] "Lectures on the Science of Language." By Prof. Max. Müller. Vol. ii., p. 633.

The observer finds that when the Sun is in "aspect" with Mars and Jupiter the temperature of the air *rises;* and that when the Sun is in aspect with Saturn and Uranus the temperature *falls.*

"Changes of weather are closely related to changes of wind, and changes of wind to changes in the distribution of atmospheric pressure,"[6] generally speaking. The changes may be constantly observed to take place simultaneously with the varying "aspects." Northerly winds and "fine weather cumulus" prevail under Jupiter's influence; easterly winds and "showery weather cumulus" prevail under Saturn's influence. Drought takes place under the combined action of Mars and Jupiter; heavy rainfall occurs under the combined action of Venus and Saturn.

The "aspects" or configurations observed to be effective in meteorology are the differences of longitude of 30°, 36°, 45°, 60°, 72°, 90°, 120°, 135°, 144°, 150°, and 180°.

Kepler suggested, in addition to those enumerated, the following aspects:—18°, 24°, and 108°.

---

[6] "Aids to the Study and Forecast of Weather." By W. C. Lay, M.A. Published by the authority of the Meteorological Council.

CHAPTER IX.

ASTRO-METEOROLOGY—Continued.

The following tables will show the various kinds of weather observed to coincide wtth the various aspects and positions of the Sun, Moon, and planets.

I.—METEOROLOGICAL TABLE OF THE SUN.

	Spring.	Summer.	Autumn.	Winter.
The SUN with Mercury.	Generally wind and rain; ☿ retrograde, ever rain.	Variable—generally showers.	Variable—misty or rainy—and windy.	Stormy; generally rainy.
Venus.	Misty or rainy.	Thunder showers—dashing rain.	Rainy.	Fog or much rain.
Mars.	Warm air; dry.	Heat; sometimes thunderstorms.	Dry and warm air.	Mild; generally fine.
Jupiter.	Mild air; pleasant winds generally N.W.	Fine and hot; sometimes thunder.	Windy; warm air.	Mild and windy.
Saturn.	Rain or snow; heavy clouds; raw air.	Hail, rain, or thunderstorms.	Cold, rainy, and stormy	Snow, rain, or heavy gales.
Uranus.	Bleak air; changes; frosty nights.	Overcast, or sudden squalls.	Dark atmosphere; gusty and cold.	Stormy or frosty.
Neptune.	Windy; generally rainy.	Showers and wind.	Windy and rainy.	Stormy; downfall.

TABLE II.—"MUTUAL" ASPECTS.

Planets.	Spring.	Summer.	Autumn.	Winter.
Neptune and Uranus.	Windy and variable.	Wind; some rain.	Rainy and windy.	Stormy.
Neptune and Saturn.	Gales; generally rain.	Wind and rain.	Wind and rain.	Stormy.
Neptune and Jupiter.	Mild; some wind.	Warm and fine.	Fair; mild air.	Generally fair; windy.
Neptune and Mars.	Mild air; generally fair.	Fine; heat.	Generally fair and warm.	Mild, windy.
Neptune and Venus.	Rain or mist.	Showery.	Mist or rain.	Fog or rain.
Neptune and Mercury.	Windy and variable.	Windy and unsettled.	Stormy or rainy.	Stormy and wet.
Uranus and Saturn.	Squally and cold; downfall.	Windy, cool, and unsettled.	Cold, showery, & squally.	Very stormy or sharp frost.
Uranus and Jupiter.	Variable, windy.	Thundershowers.	Thunder, hail or rain.	Unsettled, cold, and windy.
Uranus and Mars.	Squally; sudden changes.	Thundershowers.	Hail and thunder.	Stormy and unsettled.
Uranus and Venus.	Cloudy, showery and cold.	Cloudy, hail or rain.	Dashing rain.	Snow or rain; fog.

TABLE II. CONTINUED.

Planets.	Spring.	Summer.	Autumn.	Winter.
Uranus and Mercury.	W i n d y , variable, and cold.	Windy; hail or rain.	High winds; hail.	Gusty and cold.
Saturn and Jupiter.	Windy, cold. and often rainy.	R a i n y ; thunder.	Wind and rain.	Turbulent air.
Saturn and Mars.	R a i n y , w i n d y ; thunder.	Thunder-storms.	Windy and unsettled.	Cold a n d windy.
Saturn and Venus.	Cold and rainy.	S u d d e n rains, cool air.	Cold and rainy.	Snow or rain; fog.
Saturn and Mercury	Wind and often rain.	Windy and unsettled.	Windy and rainy.	Snow or frost; wind.
Jupiter and Mars.	Turbulent but dry air.	Great heat and thun-der.	Windy, but warm.	Mild a n d windy.
Jupiter and Venus.	F i n e growing weather.	Heat, and pleasant air.	Clear and serene air.	Mild, yet snow may fall.
Jupiter. and Mercury.	Gusty, fine generally.	Gusty, thun-der or hail-storms.	W i n d y ; hail.	W i n d y ; hail.
Mars and Venus.	Abundant rains.	Small rain prevails.	Rainy.	Snow or rain.
Mars and Mercury.	Windy and sometimes rainy.	Thunder or hailstorms.	Wind and hail.	Snow or rain.

TABLE II. CONTINUED.

Planets.	Spring.	Summer.	Autumn.	Winter.
Venus and Mercury.	Pleasant showers, misty air.	Cloudy or rainy.	Variable; misty air.	Abundant rains; sometimes floods.

TABLE III.—PLANETS STATIONARY, OR IN THE EQUATOR OR TROPICS.

Planets.	Spring.	Summer.	Autumn.	Winter.
Neptune	Windy and rainy.	Showery.	Windy and showery.	Stormy.
Uranus.	Cold, windy, and rainy.	Cool and Showery.	Windy, cold, and rainy.	Stormy, snow or rain.
Saturn.	Windy, cold, and unsettled.	Showery and cool.	Cold, and windy.	Stormy, snow or rain.
Jupiter.	Gusty, but mild air.	Heat; some thunder.	Warm air; thunder.	Stormy, but mild.
Mars.	Unsettled, but mild.	Heat; some thunder.	Warm air; thunder.	Unsettled, gusty.
Venus.	Rainy and cold.	Showery.	Rainy and cold.	Snow or rain.
Mercury.	Windy and unsettled.	Variable; gusty.	Rainy and windy.	Stormy, snow or rain.

Certain of the various configurations and relative positions of the Sun, Moon, and planets, are found to operate with greater force than others: 1. The Solar configurations are of primary importance. 2. The position of the Moon and planets in equator or tropics, the northern tropic affecting mostly the northern hemisphere. 3. The *stationary* positions of the planets. 4. The "mutual" aspects of the planets.

The "aspects" may be divided into two classes, major and minor. The *conjunction* and *parallel declination* are of the major class, and probably more potent than any aspects. The *major* aspects are: the opposition (180°), square (90°), trine (120°), and sextile (60°). The *minor* aspects are: the semi-square (45°), sesquiquadrate (135°), quincunx (150°), biquintile (144°), quintile (72°), decile (36°), tredecile (108°), semi-sextile (30°), quindecile (24°), and vigintile (18°).

The solar aspects of the major class usually operate for a period of from two to four days—the parallels of declination always begin to act the day before they are complete. The *stationary*, equatorial, and tropical positions of the superior planets exert a disturbing influence on the weather for two or three days—in some cases, when no other positions interfere, for even seven or more days. The major aspects of the superior planets operate for about two or three days, usually. The major aspects of the inferior planets (Mercury and Venus) and the minor aspects of the superior planets (Mars, Jupiter, Saturn, Uranus, and Neptune) do not usually operate for a longer period than twenty-four hours.

CHAPTER X.

ASTRO-METEOROLOGY—Continued.

HOW TO PREDICT THE WEATHER.

"To place the forecasts of weather, even of the general weather of the coming season, on a sound and certain basis, to gain the power of foretelling a cold spring, a wet summer, or a late harvest, would be to confer an incalculable benefit upon the people of this country."—The *Times*, September, 1878.

In my "Weather Guide Book," published in the year 1864, I gave rules for foretelling the general character of the weather. Those rules (revised) are now once more presented to the consideration of the scientific world and the intelligent public.

1. Tabulate the aspects and relative positions of the Sun, Moon, and planets, in three columns, as shown in the table of phenomena on the succeeding page.

PHENOMENA, JULY, 1881.

	Solar.	Mutual.	Lunar.
1	60° ♄, 45° ♀.	♂ p.d. ♄.	
2	60° ♅.		Eq. 5^h p.m., ☌ ♅.
3		♀ 18° ♂.	
4		☿ stationary.	□ ☉ 5^h 16^m p.m.
5	18° ☿.	☿ 72° ♃. ♀ 18° ♄.	
6		♂ ☌ ♄, 120° ♅. ♀	
7		☿ p.d. ♀ [p. ♃.	
8	60° ♆.	☿ 60° ♀, p.d. ♃.	
9			S. tropic, noon.
10			
11			☍ ☉ 2^h 13^m p.m.
12	60° ♃.	♂ p.d. ♆. ♀ 24° ♄.	Perigee, 2^h a.m.
13		☿ 72° ♂ et ♆.	
14		♂ ☌ ♆.	
15	72° ♄.		Equator, noon.
16		☿ 45° ♅.	
17	Inf. ☌ ☿.		□ ☉ 5^h 33^m a.m.
18	45° ♅.	♄ 120° ♅. ☿ 45° ♀	
19		♀ □ ♅, 30° ♄.	☌ ♄ 10^h 32^m a.m.
20		☿ 72° ♄.	☌ ♂ 3^h 7^m a.m. ☌
21	72° ♆.		[♃ 5^h 44^m a.m.
22		♂ ☌ ♃ 4^h 36^m p.m.	N. tropic 10^h a.m.
23		☿ 60° ♂ et ♃ ♂ p ♃	[☌ ♀.
24	p.d. ♀.	☿ p.d. ♃. ♀ 30° ♆.	
25		♀ 36° ♄.	☌ ☿ 4^h 42^m a.m.
26	45° ♀.	♀ 24° ♂.	☌ ☉ 5^h 19^m a.m.
27	72° ♃.	[♀.	
28	36° ♅.	☿ stat., p. ♂, 30°	
29			Eq. 11^h p.m. ☌ ♅.
30	p.d. ☿ et ♂.	♀ 30° ♃. 36° ♆.	
31		♄ 120° ♅.	

2. If the solar configurations with Mars and Jupiter preponderate, and especially if Jupiter be *stationary* or in mutual aspect with Mars, there will be but little rain and the temperature will be above the average.

3. If the Solar aspects with Saturn, Uranus, Venus, and Mercury (retrograde) preponderate, expect much downfall and low temperature.

4. When the Sun is in aspect with Mercury (the conjunction and parallel declination) expect rapid changes and strong wind. When, at or near the same time, Mercury is in aspect with Saturn, Uranus, or Neptune, expect gales or storms of wind. When, at or near the same time, Mercury is in aspect with Mars or Jupiter, expect elevation of temperature, and hail or thunderstorms.

5. When Mars or Jupiter is in equator or at extreme north declination expect high temperature, with occasional gusts of wind. When Saturn, Uranus, or Neptune is in such position, expect stormy weather. When Venus is so placed, expect much rain and some wind.

6. When there are several mutual aspects of the planets (within a few days) expect great atmospheric disturbance ; and when two or more configurated planets are, at the same time, in *Taurus* or *Scorpio*, or near the equator or tropic, expect volcanic action and great storms.

7. When at the equinox or solstice, either Uranus or Saturn is in the lower meridian and configurated with the Sun, expect a cold and unsettled season ; when Mars or Jupiter is so situated expect a fine and propitious season ; when Venus is so situated expect much rain.

In *Urania* for January, 1880 (in my article on " The Weather and its Prediction "), several instances were given of the angular positions of the planets Saturn and Uranus at the equinoxes and solstices being followed by unusually cold and wet seasons.

If we refer to the map of the heavens for the winter solstice, 1881, given at page 15, we shall find that Uranus is in the lower meridian, and, therefore, we may anticipate a *very cold winter* in 1881-2; but as Mars is in the ascendant and in opposition to the Moon, there will be mild weather at intervals. The *coldest* periods of the ensuing winter will probably be from January 16th to 26th; February 16th to 22nd ; 25th and 26th; and March 4th to 7th, 1882. The *mildest* periods: January 5th, 9th to 11th ; February 3rd to 6th, 15th ; March 11th to 13th, 26th and 31st. The *stormy* periods : December 19th, 23rd and 27th ; January 6th, 18th, 23rd to 25th ; February 2nd, 16th and 26th; March 4th, 7th, 18th, 20th, and 31st. The periods of *much downfall:* December 23rd, 1881; January 4th, 25th, and 31st ; February 17th. 20th to 22nd, and 26th; March 5th, 17th, 21st, and 29th, 1882.

A glance at the *phenomena*, of July, 1881, and a comparison of them with the meteorological tables given in chapter IX., will explain the extremes of temperature and sudden variations of weather that characterised that month. [Comet *b*, 1881, was then visible]. On the 5th of July, 1881, the shade temperature reached 93° in the South of London. The influence then operating was VENUS *par. dec.* JUPITER. This was followed, in the early morning of the 6th, by a violent thunderstorm which prevailed over the whole of England and Wales and part of Ireland, doing an immense amount of damage. At 6^h 56·8^m a m. on the 6th of July, the *conjunction* of MARS and SATURN took place in *Taurus* 10° 24′ 15″. An extremely low temperature followed. Again, as the influence of the SUN in *sextile* aspect (60°) with

JUPITER (on the 12th inst.) came into operation, the temperature rose again (on the 11th inst) until it reached 97·1°, in the shade, at Greenwich Observatory, on the 15th inst. This extreme heat was partly due to the approaching *conjunction* of MARS and JUPITER in *Taurus* 21° 47′ 58″. When this conjunction was complete (on the 22nd inst.), and the influence of the SUN *par. dec.* VENUS came into play, the temperature again fell and rainy weather followed. During the closing days of July, 1881, Saturn was in *trine* (a very rare aspect) with Uranus, and the weather, accordingly, became unsettled and very cool.

In case a doubt should arise as to the storm of the 6th of July, 1881, being due to the conjunction of Mars and Saturn, we will refer to the conjunctions of those planets which have occurred during the last few years. First, let us note that Zadkiel foretold that this conjunction would be attended by " a most violent storm;" and that the Meteorologic Office issued no warning whatever of it.

There was a conjunction of Mars and Saturn on the 30th of June, 1879, at 7ʰ 38ᵐ p.m., in *Aries* 15° 8′ 38″. On that night, a very sudden and rapid fall of the barometer took place, and a heavy gale set in, on the south-west coasts of England, resulting in sad disasters to shipping on the Cornish coast. In this case also the Government meteorologists issued no warning of a coming storm, but announced that the weather was " improving;" whereas Zadkiel had published, nine months beforehand, his forecast of " a great storm on the 30th of June."

On the 3rd of November, 1877, Mars and Saturn were conjoined, at 11ʰ 28.3ᵐ p.m., in *Pisces* 13° 45′ 21″. On that very night there was a great storm at Constantinople, and rough weather on our coasts. Zadkiel's forecast was: "November 3rd to 5th, a stormy and rainy period."

On the 22nd of November, 1875, Mars was in conjunction with Saturn (and in opposition to Uranus). Very stormy weather, with sleet and rain, prevailed in the north of England. Zadkiel foretold storms at this period. One more instance may be recalled, and it is a notable one. On the 11th of September, 1861, Mars was in conjunction with Saturn. Zadkiel's forecast was: September, 1861, "*a month remarkable for tempests*—11th *storms*, much rain—lo! I have warned you!" On the 10th inst. the "Great Eastern" sailed from Liverpool, no warning of an impending storm having been sent to the captain from the Meteorologic Office. On the 11th, the Great Eastern was nearly destroyed by a fearful storm. Coincidences such as these might be multiplied. Unfortunately, the Government meteorologists treat them with ridicule.

"We know nothing of physical causes except by observing instances of what appear to be invariable and necessary sequence. After a certain amount of experience we assume the invariability and the necessity; and we do so most readily when our set of experiences is backed up and supported by other sets of experiences. Thus watching for coincidences is a necessary process of scientific discovery."[1]

Astronomers would find it more profitable and more commendable to watch for such coincidences and sequences as the foregoing, than to confine their observations of the conjunctions of Mars and Saturn to the comparatively puerile amusement of "testing photometrically and also photographically the lustre of the conjoined planets."

---

[1] *Intellectual Observer*, September, 1864.

The sudden fall of temperature and heavy rainstorms of the 6th of June, 1881, after a period of intense heat, coincided with the *conjunction* of VENUS with SATURN.

The fearful snowstorm of January 18th, 1881, was due to the Sun 135° Uranus, Venus par. dec. and opposition Uranus, and Venus par. dec. Saturn.

The succession of gales which prevailed from the 7th to the 10th of January, 1866, and culminated in that terrible storm that wrecked the ill-fated " London," destroying 220 lives, coincided with the opposition of Mars and Uranus (from the tropics), and the Sun in par. dec. with Mercury.

The fearful gale of October 25th, 1859, which wrecked the " Royal Charter," coincided with the opposition of Mars and Neptune, both planets being very near the Equator.

The great Crimean hurricane of the 14th of November, 1854, coincided with the opposition of Venus and Uranus, and the parallel declination of Mercury with Jupiter and Saturn.

The late Admiral FitzRoy, in his " Weather-Book " (pp. 257-64) gave an account of the terrible tempest at Barbadoes on the night of August 10-11, 1831. That tempest destroyed five thousand lives. The whole face of the country was laid waste; no sign of vegetation was apparent. The planets Mars and Saturn were then in parallel declination, and on the 12th were in conjunction. The Sun was in opposition with Jupiter on the 10th. Zadkiel, writing of this conjunction said: " There are now (August, 1831) three planets, namely Saturn, Mars, and Mercury conjoined in the first degrees of *Virgo. Storms, wrecks,* and *violent convulsions* of nature will assuredly follow! This assemblage of the heavenly bodies always foreshows direful events. On the 11th expect stormy and bad harvest weather."

On February 28th and March 1st, 1818, the "Magicienne," frigate, was lying at Mauritius, moored in the harbour of Port Louis: and on that occasion, this frigate and forty other vessels were driven on shore, or were sunk. The barometer sunk lower than ever was known, and most of those who observed it were unable to account for the notice it gave in so extraordinary a manner. On the 27th of February, the Sun was in conjunction with Saturn; and Jupiter was in parallel declination with Uranus. On the 3rd of March, Mars was in opposition with Uranus, and Venus was in conjunction with Saturn.

In like manner numerous other instances of the coincidence of violent storms with conjunctions, etc., of the Sun and the larger planets, could be cited, did space permit. In fact, we may aver that there never happened either an extraordinary storm or violent convulsion of nature without the coincidence of violent aspects of the heavenly bodies, and *vice versâ*.

The changes of temperature which precede rain are chiefly due to planetary action; if they were due solely to solar action we should always have the same temperature, in the same place, on the corresponding day of every year. When Venus is in aspect with the Sun, Mars, Saturn, Uranus, or Neptune, a fall of rain coincides or immediately follows; for Venus seems to exert a pluvial influence.

The *amount* of rainfall in any locality depends on several circumstances. Sir John Herschel says: "The most influential being its proximity to large bodies of heated water, such a prevalent direction of the wind as shall not drift the vapour away from it, and the absence of any lofty mountains in the direction of the moist wind to act as a barrier by causing its deposition on them. As we recede from the sea into the interior of great continents rain becomes rare, especially if the soil be sandy. The west coasts of England and Ireland receive with the west and south-west winds, which generally prevail, the vapour of the Gulf Stream. In consequence, the annual fall of rain is not only much greater than on the eastern and southern coasts, but in one district, that of the Lakes of Cumberland, is quite enormous. The annual fall at Seathwaite, in Borrowdale, amounted to no less than 141·54 inches on an average of three years, while that in London is only 23½. Rain, except in the tropical regions, is, perhaps, the most irregular of all meteorological phenomena, both in respect of the frequency of its occurrence, and in the quantity which falls in a given time."

Boerhave thought that planetary action might very probably contribute to unite the primary particles of water floating separately in the atmosphere, and thus occasion rain, snow, and hail. " The generation of hail seems to depend on a very sudden introduction of an extremely cold current of air into the bosom of a quiescent, nearly saturated mass "—Sir John Herschel says.[2] Large masses of ice have occasionally fallen, and Sir John Herschel thought that Professor Tyndall's experiments on the reuniting of broken ice by " regelation," or a sort of welding, fully explain the formation of large masses of ice of irregular forms in aerial conflict. " Great hailstorms are often preceded by a loud clattering and clashing sound, indicating the hurtling together of masses of ice in the air." On the 8th of May, 1832, a mass of ice fell in Hungary a yard in length and nearly two feet in thickness ; the Sun was that day in square aspect (90°) with Uranus. Hailstorms often coincide with the major aspects of Mercury

---

[2] " Meteorology." By Sir John Herschel, Bart., K.H.

with Jupiter formed near the time of Solar aspects with Saturn or Uranus.

In Mr. Glaisher's translation of Flammarion's *L'Atmosphère* is the following paragraph :—

" Hail occurs during a thunderstorm when the temperature is very high upon the surface of the ground, but decreases rapidly with elevation. This rapid decrease is the principal element in the formation of hail, and it has been known to be as much as 1° in a little more than 100 feet. What then takes place in the region of the clouds ? Those above, from 10 to 20 or 25 thousand feet high, contain, the highest of them, ice at about 30° Fahrenheit, the lowest of them vesicular water at about zero Fahrenheit. The lower clouds contain vesicular water above 32°. As a rule these clouds travel in different directions, and hail is formed when there is a collision and admixture of winds, currents, and clouds, the temperatures of which are different. The vapour which then resolves itself into rain freezes instantaneously in so low a temperature."

Now this is to the last degree compatible with planetary action; for the sudden lowering of the temperature of the air caused by the operation of Saturn's or Uranus's influence (as, for example, May 8th, 1832, ☉ □ ♅), or the struggle between contending influences when Jupiter and Saturn are operating together, will account for the rapid formation of hail.

On the 6th-7th of May, 1862, a hurricane of wind, accompanied by vivid lightning, passed over London, the Isle of Wight, Nottingham, Derbyshire, Leicestershire, Yorkshire, and Lincolnshire. Large hailstones fell, some of them three inches long, two inches wide, and half-an-inch thick.[3] At Whittlesea the hailstones were

---

[3] See Mr. Lowe's description of this storm, in the *Intellectual Observer*, July, 1862.

from four to five inches, and at Leeds seven inches in circumference. On the 6th of May, the Sun was in conjunction with Mercury and *trine* (120°) with Saturn; and Mercury was in trine with Jupiter on the 7th. Here we find *opposing* influences operating together, and a violent storm accompanied with hail is the result.

The great thunderstorm of August 23rd-24th, 1855, resulted from the Sun being in opposition with Jupiter and in sextile with Saturn, Jupiter having the trine with Saturn; Mars in par. dec. Saturn; Mercury in opposition with Jupiter and sextile with Saturn.

The thunderstorm of September 3rd, 1841, which visited at the same time London, Paris, Rouen, Magney, Lille, and Evereux, resulted from the Sun being in square aspect with Jupiter; Saturn being *stationary* at the same time and within 1° of the par. dec. of both Mars and Jupiter.

In August, 1881, the harvest has suffered greatly from the excessive rainfall. In this month we find many pluvial aspects. Saturn *stationary* on the 25th inst., coinciding with Mars in square with Uranus, and Mercury in square with Jupiter, produced violent storms of wind, rain, and hail, the country in many parts being flooded.

Cardan observed that "whenever Saturn is joined with the Sun, heat is remitted, and cold increased; which alone may be a sufficient testimony to the truth of astrology." Three centuries of observation, since Cardan wrote, have served to prove the truth of this aphorism.

In our variable climate, next to the forecasting of periods of great storms, the successful prediction of periods of *fine* weather forms the best test of the truth of a system of predictive meteorology.

CHAPTER XI.

EARTHQUAKES.

"Possessor of the ocean's gloomy depth,
Ground of the sea, earth's bourn, and source of all!
Shaking prolific Ceres' sacred seat,
When in the deep recesses of· thy reign,
The madding blasts are by thy power confined;
But oh! the EARTHQUAKE'S dreadful force forefend!"
ORPHIC HYMN.

ARISTOTLE in his Μετεωρολογικά, published about 300 B.C., treated of meteors, water, air, and *earthquakes.* He observed that earthquakes chiefly occur "about the Hellespont, Achaia, Sicily, and Eubœa." He also placed on record the fact that "it sometimes happens that there is an earthquake about the eclipses of the Moon."[1] M. Barthélémy Saint Hilaire, commenting on this observation, says, in a foot-note: "Pendant lés eclipses de lune," c'est là une coïncidence toute fortuite; mais les deux phénomènes n'ont aucun rapport. The clever French translator and author here asserts more than he knows. An unprejudiced comparison of the dates of great earthquakes with those of eclipses of the Sun and Moon, shows very striking coincidences so frequently repeated as to lead to the conclusion that they are not "fortuitous" and may have some "rapport." It is the *planetary positions* at eclipses that

---

[1] "The Works of Aristotle," translated from the Greek, by Thomas Taylor. Vol. V. "On the Heavens, Meteors," etc. Book II., p. 528.

are the causes (or the "signs") of earthquakes. It is generally found that the shock is felt at those places where Jupiter or Saturn is angular at the moment of greatest eclipse; and that the shock usually takes place when Jupiter or Saturn (as the case may be) retrogrades over its own place at the eclipse, or when it arrives at the longitude of the eclipse or the square or opposition thereof.

The late Commander Morrison, R.N., published in the year 1834 the following observations on and rules for foretelling earthquakes:—

"1. Earthquakes generally follow close on the heels of eclipses.

"2. At the period of the earthquake, many *aspects* will be found between the planets in the heavens; also, as regards the places of the planets at the preceding eclipse, but chiefly the places of the Sun and Moon.

"3. Earthquakes happen more frequently when there are planets—especially Uranus, Saturn, Jupiter, and Mars—in the signs *Taurus* and *Scorpio*.

"4. If there have been no recent eclipse of the Moon, within a month, look to the last eclipse of the Sun.[2]

"5. The planet JUPITER, in aspect with Venus or Mercury, more especially the conjunction, opposition, and parallel declination, has a powerful influence in causing earthquakes—especially when in *Taurus* or *Scorpio*.

"6. If no eclipse have taken place within three months, look to the planets' places at the last new or full moon of the quarter, *i.e.* the lunation nearest to the Sun's crossing the equator or tropic.

---

[2] Earthquakes mostly occur in places where Jupiter or Saturn is in the meridian at the eclipse.

" 7. Earthquakes generally happen when there are several planets in or near the tropics or equator.

" 8. Earthquakes may always be expected near the perihelion of great comets, and when they approach within the orbits of the planets Uranus and Saturn.

" 9. Let all, or as many as possible, of these circumstances be combined before any very extensive earthquakes be predicted."

At the earthquake of Santa Martha, which coincided with a great eruption of Mount Vesuvius, on the 22nd of May, 1834, the geocentric longitudes and declinations of the Sun, Moon, and Planets, were:—

	☉	☽	☿	♀	♂	♃	♄	♅
Long.	0♊51	24♏39	18♉20	20♊12	9♈58	21♉30	4♎10 R	26♒30
Dec.	20 N 21	16 s 15	16 N 10	23 N 56	2 N 41	17 N 21	0 N 43	13 s 22

At the *lunation*[3] (full Moon) nearest the vernal equinox, the longitudes and declinations of the Sun, Moon, and planets, were:—

	☉	☽	☿	♀	♂	♃	♄	♅
Long.	4♈7	1♎7	10♈13 R	8♈27 25♒12	8♉0	7♎40 R	24♒49	
Dec.	1 N 38	2 N 56	7 N 10	2 N 11 14 s 12	13 N 19	0 s 35	13 s 55	

At the time of the *earthquake* Saturn was in transit over the exact place of the Moon at the *lunation*; and Mars had attained the declination of the Moon at the lunation. Moreover, at the earthquake, Saturn was within 1° of arc in declination; Jupiter, nearly conjoined with Mercury, was in *Taurus*, and the Moon (in *Scorpio*) was in opposition with Mercury and Jupiter.

---

[3] March 25th, 1834, 6ʰ 13ᵐ a.m.

On the 15th of July, 1853, a fearful earthquake happened at Cumana, at $2^h 15^m$ p.m. (mean time there), by which thousands of lives were lost. On the 6th of June, 1853, the Sun was eclipsed. The longitudes of the Sun, Moon, and planets at the moment of ecliptic conjunction of the luminaries, were:—

☉ and ☽	☿	♀	♂	♃	♄	♅	♆
15 ♊ 56	7 ♊ 42	22 ♊ 23	17 ♉ 52	20 ♐ 6 ʀ	24 ♉ 30	10 ♉ 37	13 ♓ 42

The longitudes of the heavenly bodies at the time of the *earthquake* were:—

☉	☽	☿	♀	♂	♃	♄	♅	♆
23 ♋ 5	17 ♏ 23	19 ♌ 20	10 ♌ 9	15 ♊ 26	15 ♐ 43	28 ♉ 48	12 ♉ 9	13 ♓ 30 ʀ

On the day of the Cumana earthquake Jupiter had arrived at the *opposition* of the place of the eclipse. Mars was passing over the place of the eclipse and was in opposition to Jupiter, and the Moon was in opposition to the place of Mars at the eclipse.

Reference to p. 524 of the *Nautical Almanac* for 1853, will show that the eclipse was visible at Cumana. At the moment of ecliptic conjunction, the sign *Scorpio* 11° 23′ was in the ascendant at Cumana, Uranus, Mars, and Saturn, all three in the sign *Taurus*, were setting. At the moment of the earthquake the 27th deg. of *Scorpio* was in the ascendant and Saturn was exactly setting in *Taurus* 28° 48′. In *Zadkiel's Almanac* for 1853 (page 42), will be found the following forecast:—

"As Mars and Saturn are in *Taurus* in the precedent angle of the eclipse at Panama, I have no doubt there will be a fearful amount of earthquakes there, and all about the Isthmus of Darien, the shocks extending to

Carthagena, along the northern coast of South America,
to Honduras, California, Florida, etc., and the West
Indies. These events may be looked for (among other
periods) in July, 1853, about the 16th day."
Cumana is situate on the northern coast of South
America. The earthquake took place there at the
precise period foretold by Zadkiel.
M. Delauney, of the French Academy, has been
making some researches in this connexion. Taking for
data M. Alexis Perrey's tables from 1750 to 1842, and
noting the *maxima* of the curve obtained, he finds a
first group of *maxima* commencing in 1759, and having
a period of about 12 years; a second commencing in
1756, also with a period of 12 years; and a third and
fourth group commencing in 1756 and 1773 respectively,
and each having a period of 28 years. Now the epochs
of *maxima* of the first and second groups coincide, says
M. Delauney, with the epochs when Jupiter attains his
mean longitudes of 265° and 135°, while the epochs of
maxima of the third and fourth groups correspond to
the periods when Saturn is found at the same two
longitudes. Thus earthquakes seem to pass through a
maximum when the planets Jupiter and Saturn are in
close proximity to the mean longitudes of 265° and
135°. M. Delauney is further of opinion that this
influence is due to the passage of these two planets
through cosmic streams of meteors. He gives an
approximate table of future earthquakes, indicating
particularly the years 1886, 1891, 1898, 1900, 1912,
1919, 1927 and 1930 as likely to have numerous earthquakes.
The longitude of 265° is only 5° short of the southern
tropic (270°). That of 135° is the middle of the sign
Leo, and a planet so placed is in square aspect to
another, placed in the middle of *Taurus* or *Scorpio*.

The theory that the "influence is due to the passage of Jupiter and Saturn through cosmic streams of meteors" is untenable when we reflect that great earthquakes have frequently occurred when these and other planets have held such other positions as those before enumerated.

Dr. Goad gave a list (in his work *Astro-Meteorologica* published two centuries since) of twenty earthquakes which happened while the planet Jupiter was in *Taurus*. He adds: "I am as sure as I write that this phenomenon, as great and stupendous as it is, depends upon this celestial appearance—Venus or Mercury with Jupiter."

When the fearful earthquake happened at Lisbon, on the 1st of November, 1755, the geocentric longitude of Jupiter was 187° 31', and that of Saturn was 293° 3'; Mars was in opposition to Saturn (δ and $\hbar$ being in *tropical* signs); Venus and the Sun were nearly conjoined in *Scorpio*; and Uranus was in *Pisces* (the ruling sign of Portugal). At the time of the earthquake (between nine and ten o'clock in the forenoon), Jupiter was in opposition to the longitude (Υ 7°) in which the Moon was eclipsed on the previous 27th of March.

In *Urania* for February 1880, the writer called M. Delauney's attention to the entry of Jupiter and Saturn into *Taurus* in 1881, and said that this phenomenon would produce great earthquakes. The earthquake at Chios (Scio), in the Archipelago, on the 3rd of April, 1881, which destroyed 4,000 lives and injured 1,500 other persons, occurred just as Jupiter and Saturn entered *Taurus* (in the manner described by Ptolemy), the ruling sign of the Archipelago.

On the 12th of June, 1881, thirty-four villages and one hundred lives were destroyed by an earthquake in Armenia. Venus, Jupiter, Saturn, and Neptune were assembled in *Taurus* (the ruling sign of Armenia), within the space of eight degrees.

On the 22nd of July, 1881, when Mars was in conjunction with Jupiter in *Taurus*, violent shocks of earthquake were felt in Switzerland (the sharpest since 1854), at Lyons, and at Grenoble.

The severe earthquake at Agram on the 9th of November, 1880, coincided with the opposition of Mars and Neptune. The severest shock preceded by only two hours the exact opposition of these planets, *viz.* at 10^h 3.7^m a.m., mean time at Agram. The longitude of Mars at the opposition was *Scorpio* 12° 46′ 45″, that of Neptune *Taurus* 12° 46′ 45″. The severest shock was felt at a quarter before eight o'clock in the morning, at which moment Uranus was exactly culminating in the 14th degree of *Virgo* (the ruling sign of Croatia), and the 27th degree of the sign *Scorpio* was in the ascendant.

On the 6th of March, 1867, there was an annular eclipse of the Sun, visible in Asia Minor. Zadkiel foretold that earthquakes in Asia Minor would follow on the heels of that eclipse. On the same day, a fearful earthquake happened at Mytelene, at six o'clock p.m. Shocks, more or less severe, were felt, on the 7th or 8th of the same month, over a great part of the Levant, and in some of the Ægean islands. Smyrna, Magnesia, Adramiti, Aivali, Gallipoli, and Constantinople were more or less shaken. Twenty thousand houses and public buildings, and more than two thousand lives were destroyed in Mytelene. At the eclipse, Saturn was retrograde in *Scorpio* and in square aspect with Jupiter in Aquarius.

The frequency of the occurrence of earthquakes when the larger planets are passing through *Taurus* or *Scorpio*, or in the equator or one of the tropics, cannot fail to strike the unprejudiced mind. These coincidences so oft repeated suggest a possible and probable connexion—in fact, that the planetary positions described are the exciting causes of earthquakes.

It would be out of place here to discuss the various theories that have been advanced as to the true cause of earthquakes. It is a singular fact that a writer in the *Quarterly Review* for July, 1881, advances the theory that the cause of earthquakes is *electrical*—the very theory advanced by the late Commander Morrison, R.N., in *Zadkiel's Almanac* for 1868, pp. 50-53.

Here is Commander Morrison's theory briefly given in his own words:—

" It is this absurd theory of a universal fire in the middle of the earth which blinds our *savans* to the true cause of earthquakes. We will here declare the *true* causes of these fearful phenomena, which have been long since seen by Stukesly. Everybody knows that if we charge a Leyden jar with electricity, that when it is *fully* charged it will discharge itself. Now electricity is almost constantly coming into the earth from the air, or going from the earth into the air; and when it *accumulates* in a given place, it will then discharge itself, as does the Leyden jar. The result is a shock, more or less severe, of earthquake. If slight, it may derange the internal structure, perhaps many miles deep, and then a *rumbling* noise is heard. If this be not heard and the shock be severe, it tears up and rives to pieces the most solid rocks, and laughs at the puny buildings of mortal man. Such was the case recently in the fair island of Mytelene.

" It very frequently occurs in the vicinity of volcanoes, both active and extinct; and in such cases the *discharge* of electricity, as we have ourselves frequently witnessed at Naples, Lipari, and Etna, is attended with thunder and lightning, *hail, rain*, and sudden gusts of wind, just similar to such phenomena in England.

"If we be right in our theory, that earthquakes are caused by *accumulations of electricity*, then it follows, that if there be a cause constantly, or nearly constantly, happening to carry off the electricity and prevent its accumulation, there will be never any very serious results from the discharge in that locality. Now, in England we have very frequent, and we may say, almost constant *rains*. These operate to carry away the electric fluid (which comes to us from the air) and conduct it into the rivers and thence into the sea. The same effects are produced by the rains and snows in Russia, Holland, and some other countries. It is chiefly in the parts of the earth subject to much heat and drought, that these dire events of sudden and fearful earthquakes or discharges of electricity take place. If any shocks occur, as they sometimes do in England, it will always be found that they succeed long and extraordinary droughts, which are, fortunately for us, very rare. The character of electricity is very peculiar. It acts always instantaneously. It flies across the Atlantic in something less than a hundredth of a second. A few years since there was a violent shock of earthquake in the West Indies and the Arkansas (fully 2,000 miles asunder) at the same instant of time ! In fine, earthquakes have all the characteristics of electric discharges. There is nothing absurd or improbable in their being produced thereby. Whereas the theory that supposes a vast internal fire raging at all times in the centre of the earth, has nothing to commend it to our notice and has much that shocks our common sense."

The writer in the *Quarterly Review* says: " Considering the irresistible force, the unmeasured rapidity, the quick repetition and long duration of the shocks, what known agent in nature, we would ask, except *electricity* is capable of producing at the same time such singular effects in the sea and such tremendous results on land ?"

Drs. Priestly and Stukesly attributed earthquakes to *electricity*, but they were unable to explain how they were brought about. Planetary influence, alone, can supply the clue.

On the 17th of May, 1882, there will be a total eclipse of the Sun in the sign *Taurus* 26° 15'. The luminaries will then be but 9° 10' separated from the conjunction with Saturn. On the 8th of September, 1882, Saturn will be *stationary* in 26° 11' of *Taurus*, *i.e.* in the very place of the eclipse. Earthquakes in Asia Minor, in the Archipeiago, and in Persia ,may be expected about the 17th of May and the 8th of September, 1882.

CHAPTER XII.

GENETHLIALOGY.

" Ye stars! which are the poetry of heaven,
If in your bright leaves we would read the fate
Of men and empires—'tis to be forgiven
That, in our aspirations to be great,
Our destinies o'erleap their mortal state,
And claim a kindred with you ; for ye are
A beauty and a mystery, and create
In us such love and reverence from afar,
That fortune, fame, power, life, have named
themselves a star."—BYRON.

PLANETARY influence upon individuals is pretty generally discredited and ridiculed. Even many who admit that evidence appears to point very strongly to planetary influence on the atmosphere and weather-changes, thereby affecting mankind indirectly at least, draw the line at nativity-casting. This cannot be wondered at, for *primâ facie* it does seem absurd that the accidental positions of the heavenly bodies at the moment of birth of a child should influence his health, fortunes, affections, and disposition, throughout his earthly career.

Lord Bacon, while advocating an *Astrologia Sana*, once regarded as " an idle figment " the " doctrine of

F

horoscopes and houses," and the calculation of nativities
as altogether superstitious (lib. vi., cap. 3); but he
afterwards averred that "astrological knowledge gives
us some apt distinctions of men's dispositions, according to the predominance of the planets" (lib. vii.,
cap. 3).

Kepler pursued genethliacal astrology.   We are in
formed that "though on behalf of the world he worked
at astronomy, for his own daily bread he was in the
employ of astrology, making almanacks and drawing
horoscopes that he might live."[1]   Were there no truth
whatever in judicial astrology, and were the practice of
it a fraud as its enemies aver, so honest a man as Kepler
could never have practised it.   He cannot be classed as
"either a fool or a knave," and we may rest assured
that were astrology false he would have discovered and
proclaimed it to be so.   To so clever a man as Kepler
other means of earning his daily bread were open, and
the insinuation of Mr. Blake that Kepler only practised
astrology from necessity, and at the cost of his honesty
and dignity, is unfounded and discreditable.   In their
anxiety to prove (without examination be it remembered) astrology to be false, its enemies appear to be
careless of whose reputation may suffer through their
hasty condemnation of the art.

As Stahl said: "It is foolish to make that a matter
of discussion which any one may decide by experiment."

Let us leave the discussion forum and pass into the
laboratory.   Before commencing our experiments, how-

ever, we must make ourselves thoroughly acquainted
with the symbols of our science:—

☉	The Sun.	♈	Aries.	☌	Conjunction.
☽	The Moon.	♉	Taurus.	⚹	Sextile.
☿	Mercury.	♊	Gemini.	□	Quadrature.
♀	Venus.	♋	Cancer.	△	Trine.
♂	Mars.	♌	Leo.	☍	Opposition.
♃	Jupiter.	♍	Virgo.	☊	Ascending node.
♄	Saturn.	♎	Libra.	☋	Descending node.
♅	Uranus.	♏	Scorpio.	°	Degrees.
♆	Neptune.	♐	Sagittarius.	'	Minutes of Arc.
		♑	Capricornus.	"	Seconds of Arc.
		♒	Aquarius.		
		♓	Pisces.		

The division of the heavens into twelve "houses" has
been already described at pp. 10-11. The process employed in casting the horoscope will be delineated in the
next chapter.

CHAPTER XIII.
ON CASTING THE HOROSCOPE.

" To whom the heavens, in thy nativity,
Adjudged an olive branch and laurel crown,
As likely to be blest in peace and war."
SHAKESPEARE.

THE HOROSCOPE must be cast for the true moment of birth—*i.e.*, when the child first cries. When the birth-place is remote from London the Greenwich mean time (which is now kept, under the name of " railway time," throughout England and Wales) must be corrected to *local* time by *adding* four minutes for every degree of *east* longitude the birthplace may happen to be from Greenwich, or *subtracting* four minutes for every degree of *west* longitude, from the Greenwich mean time.

1. The *Nautical Almanac*[1] or *Zadkiel's Ephemeris* for the year of birth will show the amount of *sidereal time* at the mean noon *preceding* the time of birth. Write this down. *Add* to this the number of hours and minutes that have elapsed since the preceding noon. *Add* also the correction for the difference between mean and sidereal time at the rate of 9.86^s per hour. If the sum exceed 24^h reject this amount. Then, the sum is the *Right-Ascension* of the *meridian* at the moment of birth. Having drawn a circle in the manner before described,

2. With a " Table of Houses " for the nearest lati-tude, find the nearest *right-ascension* to that which you have obtained, and mark the values therein given on the cusps (beginnings) of the several houses, the numbers of which are given at the heads of the columns respectively, entering on the opposite houses the same degrees of the opposite signs.

---

[1] The *Nautical Almanac* does not give the geocentric longitudes of the planets, but it gives their apparent R.A. and dec.

3. Reduce the *longitudes* of the Sun, Moon, and planets, by proportion, from the *Ephemeris*, to the moment of birth (corrected to Greenwich mean time); and place their symbols in the proper houses (which may be readily done by noting that the degrees pass over the cusps of the houses from left to right).

4. Reduce the *declinations* of the Sun, Moon, and planets, by proportion, from the *Ephemeris* (or from a table of declinations), to the moment of birth.

EXAMPLE:—Let it be required to compute the horoscope of Her Majesty Queen Victoria, Empress of India. The official bulletin published in the *Courier* stated that her birth took place at 4^h 15^m am., of May 24th, 1819, at Kensington Palace.

The following is the process:—

	h.	m.	s.
Sidereal time at at noon of May 23rd, 1819	4	1	1·2
Add time elapsed since......................	16	15	0
Add diff. mean and sidereal time......	0	2	40·2
R.A. of meridian at birth .... =	20	18	41·4

In the Table of Houses for London we find that the nearest R.A. to this is 20^h 17^m 3^s, and that this gives Gemini $5\frac{1}{2}°$ in the ascendant. If worked out by "Formula 1" given in the "Text-Book of Astrology," vol. i., p. 241, the ascendant is ♊ 5° 58′ 7″. Mark this on the line representing the ascendant, and its opposite ♐ 5° 58′ 7″ on the cusp of the descendant. Referring again to the tables we shall find that we have to place on the cusp of the second house ♋ 26°; on the third, ♋ 13°; on the fourth, Leo 2°; on the fifth, ♌ 27°; and on the sixth, ♎ 7°; and the opposite signs on the opposite houses. Four signs are found "intercepted," ♍ in the fifth house, ♓ in the eleventh, ♏ in the sixth, and ♉ in the twelfth. From either the *Nautical Almanac* or *White's Ephemeris* for 1819,

the longitudes of the Sun, Moon, and planets will now
have to be computed, by proportion, for the moment
of birth. The Sun at noon of May 23rd, was in II
1° 27′ 53″, and on May 24th, II 2° 25′ 32″; his motion
in 24$^\text{h}$ was 57′ 39″. By proportion: As 24$^\text{h}$ to 16$\frac{1}{2}^\text{h}$ so
57′ 39″ to 39′ 4″, which added to the Sun's long. at
noon of May 23rd gives II 2° 6′ 57″ as his long. at the
moment of birth. In like manner, the longitudes of the
Moon and planets will have to be computed. Then
their symbols and longitudes will have to be entered in
the proper signs and houses, and the horoscope will be
complete:—

FIG. 4.

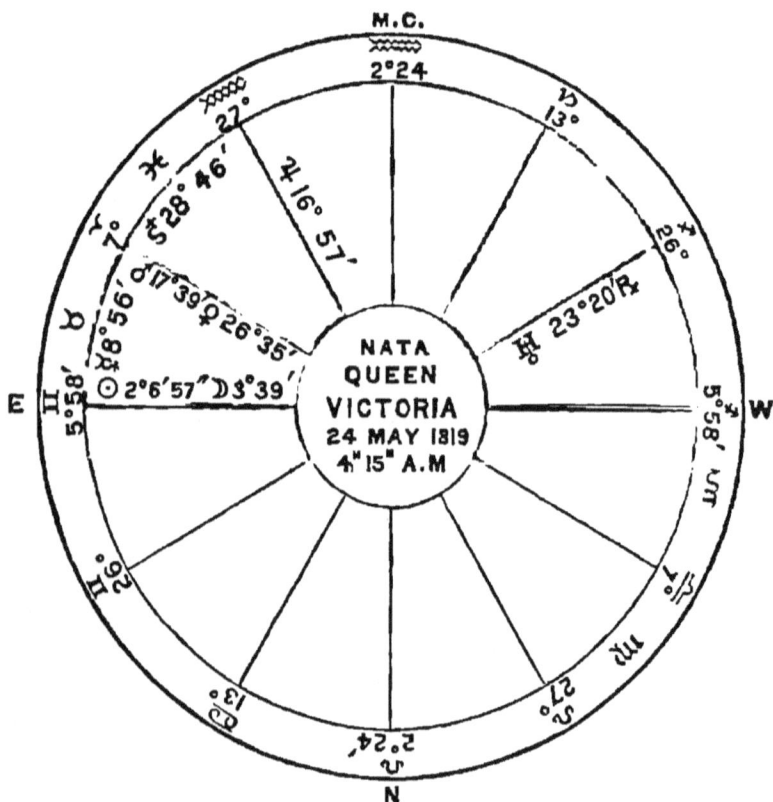

	Lat.		Dec.		R.A.		M.D.		S. Arc.	
	°	′	°	′	°	′	°	′	°	′
☉	*	*	20	36½ N	60	1	115	21	118	13
☽	3	34 N	24	23 N	60	52	116	12	124	25
☿	3	35 s	11	5 N	37	42	93	2	104	16
♀	1	56 s	8	28 N	25	21	80	41	100	47
♂	1	10 s	5	51 N	16	42	72	2	97	24
♃	0	39 s	16	23 s	319	37	14	57	68	18
♄	2	6 s	2	25 s	359	42	55	2	86	57
♅	0	8 s	23	26 s	262	43	41	57	56	58

Those of my readers who are conversant with trigonometry will be able to make these calculations quickly and easily. Those who are not so far advanced need not attempt any great exactness at first, the primary object being to comprehend the method of computation, which is really very simple.

It will be observed that the Sun and Moon had just risen at the birth of Her Majesty; Jupiter was in the midheaven and in sextile aspect with Mars. These are very fortunate positions, indeed, and they were the (astrological) signs of the accession of our good Queen to the throne and of her brilliant and fortunate reign. The lines of Shakespeare, quoted at the head of this chapter, are peculiarly appropriate to Her Majesty's nativity.

---

[2] The lat. of the birthplace is 51° 30′ 25″N., the long. 51ˢw. The R.A. of the M.C. in arc is 304° 40′ 22″. The declination of the ascending degree is 19° 38′ 51″N.

The latitudes of the Moon and planets are computed from the Nautical Almanac; their declination, right ascension and semiarcs, from the tables given in the "Text-Book of Astrology."

CHAPTER XIV.

ON FORMING A GENERAL JUDGMENT
CONCERNING A NATIVITY.

" What potent spirit guides the raptured eye
To pierce the shades of dim futurity?"

CAMPBELL.

THE horoscope being cast, the next problem is to decipher its signification.

The primary question is, necessarily, the probable duration of life, which, as Ptolemy remarked, is distinct from the question of rearing.

A healthy and well-developed child, born at full term, has, generally speaking, a good chance of life, barring accidents.

A puny, weakly, sickly child, born prematurely, has but a poor chance of life. The reader will say it needs no astrological knowledge to teach or confirm this.

However, many apparently healthy children, with every circumstance in their favour, succumb to infantile disease even before the period of dentition. On the other hand, many puny and sickly children have been reared by careful nursing and judicious treatment. It is these cases that form a fair and proper basis for testing the truth of astrological aphorisms as to the probable duration of life.

In cases of *congenital* disease proving fatal within a very short period after birth, it cannot fairly be ex. pected that the horoscope should in every case indicate the early death of the infant. It is in the case of disease (or accident) coming on *after* birth, that we should expect the horoscope to pre-signify the evil.

Claudius Ptolemy says:—

"If either of the two luminaries be in an *angle,* and one of the malefics [Mars and Saturn, to which we may now add Uranus] be either in *conjunction* with that luminary, or else *distant in longitude from each luminary in an exactly equal space,* so as to form the point of junction of two equal sides of a triangle, of which two sides the luminaries form the extremities; while, at the same time, *no benefic* (Venus or Jupiter) *planet may partake in the configuration,* and while the rulers of the luminaries may be also posited in places belonging to the malefics, *the child then born will not be susceptible of nurture, but will immediately perish.*"

Here is an aphorism that can be readily tested if we can only procure sufficient *data* beyond dispute. Unfortunately, accoucheurs do not usually record the exact moment of birth of the children they bring into the world. However, we have one such case given in a medical journal:—

CASE 1.—The *Medical Press and Circular,* December 26th, 1877, gave the moment of birth (*viz.* 3ʰ 40ᵐ p.m. of November 20th, 1877) of a child born at the Rotunda Lying-In Hospital, Dublin. The mother, who was only nineteen years old, suffered with puerperal convulsions. The child (female) was "very puny and feeble," and "very great difficulty was experienced in establishing respiration, artificial respiration being kept up for nearly five hours before natural respiration was fully established;" but the child lived "only for about twenty hours."

The latitude of Dublin is 53° 23' N., and the longitude 6° 20' W. of Greenwich. According to the *Nautical Almanac* for 1877, the sidereal time at noon of November 20th was .... 15ʰ 58ᵐ 25·28ˢ

Add correction for long. of Dublin ... 0 0 4·77

Sidereal time at Dublin............= 15 58 30·05
Add time of birth 3 40 0
Add diff. mean and sidereal time ...... 0 0 36·14

R.A. of M.C. at birth=294° 46' 33" or 19 39 6·19

As the horoscope is given in the "Text-Book of Astrology," vol. i., p. 54, there is no need to reproduce it here. Reference to it will show the ascendant to be ♉ 24° 44'; the Moon rising in ♉ 25° 18', in opposition to the Sun in the western *angle*, in ♏ 28° 29' 15", and in *square* aspect to the planet Uranus in ♌ 29° 23'. The Sun (hyleg) has no assistance from either Venus or Jupiter; the Moon has the par. dec. of Jupiter (and this accounts for the recovery of the mother).

In this horoscope the evil Uranus is found *distant from each luminary in an exactly equal space* (90°), and the Sun is an *angle*. Thus do we find Ptolemy's aphorism borne out in a very remarkable manner, for the afflicting planet is one that was unknown to Ptolemy. Everything was done for the child that surgical skill could suggest, yet, the heavens at birth being constituted in the manner described by Ptolemy, the child was "insusceptible of nurture" and did "immediately perish."

Case 2.—A female infant was born at 3ʰ 20ᵐ p.m. of February 1st, 1874, in lat. 54° 54' N., and long. 1° 24' W. The R.A. of the meridian was 0ʰ 6ᵐ 38.2ˢ. The ascendant was ♋ 29°. The Moon was in the ascendant, in ♌ 14° 12', in *conjunction* with Uranus, in ♌ 8° 24', retrograde, and in opposition to the Sun, in ♒

12° 40'. The Sun was in *conjunction* with Mercury, in ♒ 11° 15', with Venus, in ♒ 7° 21', and with Saturn, in ♒ 5° 59'. In this case, the Sun (hyleg) was in conjunction with one of the benefics (Venus), but the good influence was vitiated by her (Venus's) conjunction with Saturn and her opposition with Uranus. The child died on the 19th of February, 1874, of marasmus (congenital). In this horoscope the affliction is by *conjunction* of the malefics with the Sun and Moon; and, therefore, it is in accordance with Ptolemy's aphorism.

CASE 3.—A male child, "J. Y.," was born on the 26th of October, 1874, at 6ʰ 25ᵐ a.m., in lat. 54° 54' N., long. 1° 25' W. R.A. of meridian 8ʰ 42ᵐ 51.58ˢ or 130° 42' 54". Ascendant ♎ 27°. The Sun in the ascendant, in ♏ 2° 41' 28" in *square* aspect with Saturn (in the lower meridian), in ♒ 7° 49'. The Moon in the western angle, in ♉ 16° 24', in *square* with Saturn, and in *square* with Uranus, also, in ♌ 15° 0'. In this case the *accoucheur* had to keep up artificial respiration for twenty minutes before natural respiration was fully established. The child was well developed and apparently healthy, but he only lived 35 hours. The mother died (of pneumonia) two days after delivery. In this horoscope both Saturn and Uranus attack the Sun and Moon from *angles*, and the luminaries have no assistance from either of the benefic planets.

The writer has examined many hundreds of horoscopes, and the foregoing are the only instances in which he has found such positions as those described by Ptolemy in his aphorism.

How different was the fate of Goethe, who was born under happier configurations! In his "Autobiography" the poet says:—

"On the 28th of August, 1749, at mid-day, as the

clock struck twelve, I came into the world, at Frankfort-on-the-Maine. My horoscope was propitious; the Sun stood in the sign of the Virgin, and had culminated for the day; Jupiter and Venus looked on him with a friendly eye, and Mercury not adversely; while Saturn and Mars kept themselves indifferent; the Moon alone, just full, exerted the power of her reflection all the more, as she had reached her planetary hour. She opposed herself, therefore, to my birth, which could not be accomplished until this hour was passed.

"These good aspects, which the astrologers managed, subsequently, to reckon very auspicious for me, may have been the causes of my preservation; for through the unskilfulness of the midwife, I came into the world as dead, and only after various efforts was I enabled to see the light."

The horoscope of Goethe is given in the "Text-Book of Astrology," vol. i., p. 56. An inspection of it will show that the Sun was in the meridian attended by Venus, and the Moon was in the lower meridian attended by Jupiter. The luminaries were unafflicted and were supported by the fortunes.

The danger from asphyxia was, no doubt, due (astrologically speaking) to the planet Saturn being in the ascendant in the hour before noon. Indeed, it is open to doubt whether the danger was due to the unskilfulness of the midwife. It will be observed that Goethe says that his preservation may have been due [under God] to the positions of the benefic planets.

Sir Isaac Newton's birth was premature, and he was so diminutive when born, and apparently so feeble, that he was not expected to live many hours. Yet he survived and exceeded even the average term of human existence. He was born " between one and two o'clock in the morning of the 25th of December (old style),

1642."[1] Neither of the luminaries was in an angle, and neither was afflicted (with the exception that the Moon had the sesquiquadrate of Uranus); the Sun having the sextile of Jupiter.

In regard to children dying in infancy from diseases which may or may not have been congenital, or from accidents; in all the horoscopes examined by him the writer has never found one that was free from severe affliction. This is a very striking fact, for it is the outcome of an experience extending over twenty-two years. Moreover, the writer is in constant correspondence with several medical gentlemen who always send him the particulars of remarkable births.

It has been stated that there are three planets—Mars, Saturn, and Uranus—of evil influence, and two—Venus and Jupiter—of benefic influence. Mercury and the Moon are convertible—*i.e.*, of good influence when with the benefic, and of evil influence when with the malefic planets. The Sun is of benefic influence, but when in evil configuration with the malefics, his good influence becomes deteriorated. This preponderance of evil influence over the good may partly account for the fact that half the children born die under ten years old. The ancients observed that Saturn's influence was particularly fatal to children, hence the fable arose that Saturn devoured his own children. Lemprière says: "Saturn always devoured his sons as soon as born. The god is generally represented as an old man ; in his left hand he holds a child, which he raises up, as if instantly to devour it."

Despite the utmost care, diseases of a dangerous nature, and accidents of a violent kind, occasionally

[1] "Memoirs of the Life, Writings, and Discoveries of Sir Isaac Newton." By Sir David Brewster, K.H.

happen to children, and place their lives in the greatest
jeopardy, if they do not actually prove fatal.   Can
astrology forewarn the loving parents of times of danger
to their children?   Does the science afford the physician
a certain prognosis?   The votaries of it will answer
these questions in the affirmative.   It is important to
know if there be any grounds for such confident
assurances; for, if so, then the world is unwittingly
neglecting priceless foreknowledge, and the refusal to
accept it becomes a blunder.

It is related in the " Life of Dryden " that he was
much addicted to judicial astrology, and that he used to
cast the horoscopes of his children.   At his request, the
birth of his son Charles was accurately noted.   Dryden
cast the horoscope.   When his lady was pretty well
recovered, Dryden informed her that the child was born
at an unlucky hour, for the Sun, Venus, and Jupiter
were all below the earth, and the ruler of the ascendant
was afflicted by the square aspect of both Mars and
Saturn.   "If he arrives at eight years," said Dryden,
" he will be in danger of a violent death on his birth-
day anniversary; if he should then escape, I see but
small hopes.   He will in his twenty-third year be again
under evil directions, and should he again escape, the
thirty third or thirty-fourth year, I fear, is——."   Here
the poet was interrupted by the grief of his lady who
could not bear to hear of so much calamity likely to
befal her son.   Just before the child entered his eighth
year, the poet was invited to the country seat of the
Earl of Berkshire (his brother-in law) at Charlton in
Wilts ; he took Charles with him.   Dryden, perhaps
through fear of being looked upon as superstitious, was
very cautious of allowing anyone to know that he was a
votary of astrology, and therefore could not excuse his
absence on his son's anniversary from a hunting party

Lord Berkshire had invited. However, before going out, he took care to set Charles a double Latin exercise, with a very strict injunction not to stir out of the room until his return. Charles was carefully doing his exercise when, as ill-fate would have it, the stag made towards the house, and the noise alarming the servants, they all hastened out to see the sport, one of them taking young Dryden by the hand and leading him out. Just as they came to the gate, the stag being at bay, made a bold rush and leaped over the court-wall, which was very low and very old, and the dogs following threw down part of the wall, ten yards in length, under which Charles lay buried. He was immediately dug out, and after languishing for six weeks in a dangerous state, he recovered. In his twenty-third year, Charles, being seized with giddiness, the heat of the day being excessive, fell from the top of an old tower belonging to the Vatican at Rome. He again partly recovered, but was ever after in a delicate state. In his thirty-third year he was, unfortunately, drowned at Windsor. Thus were the poet's calculations only too sadly verified. It is a pity that a copy of Charles Dryden's horoscope is not procurable.

To return to Ptolemy's observations. He says:—

"Of all events whatsoever which take place after birth, the most essential is the continuance of life; and as it is, of course, useless to consider, in cases wherein the life of a child does not extend to the period of one year, what other events contingent on its birth might otherwise have subsequently happened, the inquiry into the duration of life consequently takes precedence of all other questions. The discussion of this inquiry is by no means simple, nor easy of execution; it is conducted in a diversified process, by means of the governance of the ruling places."

The hylegliacal (vital) places are, according to Ptolemy,—

" The sign on the angle of the ascendant, from the fifth degree above the horizon to the twenty-fifth degree below it; the eleventh, tenth, and ninth houses; and, lastly, the seventh house." The Sun is hyleg when situated in one of the hylegliacal places; when the Sun is not so situated, but the Moon is, then she is hyleg; when neither the Sun nor the Moon happens to be in any one of those places, then the ascendant is hyleg. The Moon has always much influence over the physical faculties.

The liability to diseases and accidents is judged from the planets rising or setting, and those configurated with the *hyleg*. If the hyleg be free from affliction and supported by the benefics, and no evil planet be rising or setting, the physical constitution is said to be sound, healthy, and robust. When, however, the hyleg is afflicted and unsupported by the benefics, then early death is too likely to take place. When the Sun, Moon, and ascendant are afflicted by more than one of the malefics, a liability to accidents and a violent or sudden death is incurred; the greatest effects being connected with angular positions.

" Such parts of the signs," says Ptolemy, " as contain the afflicted part of the horizon, will show in what part of the body the misfortune will exist, whether it be a hurt, or disease, or both; and the natures of the planets, in operating the misfortune, also regulate its particular form or species. For, among the chief parts of the human body, Saturn rules the right ear, the spleen, the bladder, the phlegm, and the bones; Jupiter governs the hand, the lungs, and the arteries; Mars, the left ear, the kidneys, the veins, and the reproductive organs; the Sun rules the eyes, the brain, the heart, the nerves, and all the right side; Venus, the liver, the flesh, and the nostrils; Mercury, the speech, the understanding, the bile, and the tongue; the Moon, the palate, the throat, the stomach, the abdomen, the uterus, and all the left side.

"Blindness of one eye will ensue when the Moon may be in either the ascendant or descendant, either operating her conjunction or being at the full: it will also happen should she be configurated with the Sun in any other proportional aspect, and be at the same time connected with any of the nebulous clusters in the zodiac, such as the cloudy spot of Cancer, the Pleiades of Taurus, the arrow-head of Sagittarius, the sting of Scorpio, the parts about the mane of Leo, or the urn of Aquarius. Moreover, both eyes will be injured should the Moon be in an angle, and in her decrease, and Mars or Saturn, being matutine, ascend in succession to her. Or again, if the Sun be in angle, and these planets ascend before him, and be configurated with both the luminaries, whether the luminaries be in one and the same sign, or in opposition; provided, also, the said planets, although oriental of the Sun, be occidental of the Moon. Under these circumstances, therefore, Mars will cause blindness by a stroke or blow, or by the sword, or by burning; and, if he be configurated with Mercury, it will be effected either in a place of exercise or sport, or by the assault of robbers. Saturn, however, under the same circumstances, produces blindness by cataract, or cold, by a white film, or by other similar disorders.

"If Saturn and Mercury, in conjunction with the Sun, be in one of the before-mentioned angles, the native will have some defect in the tongue, and stammer or speak with difficulty: especially if Mercury be occidental, and both he and Saturn configurated with the

G

Moon. Should Mars, however, be found together with them, he will for the most part remove the defect in the tongue, after the Moon shall have completed her approach to him.

" Further, should the malefics be in *angles*, and the luminaries, either together or in opposition, be brought up to them; or if the malefics be brought up to the luminaries [by direction], especially when the Moon may be in her node, or in extreme latitude, or in obnoxious signs such as Aries, Taurus, Cancer, Scorpio, and Capricorn, the body will then be afflicted with excrescences, distortions, lameness, or paralysis.

" If the malefics be in *conjunction* with the luminaries, the calamity will take effect from the very moment of birth : but should they be in the midheaven, in elevation above the luminaries, or in opposition to each other, it will then arise out of some great and dangerous accident, such as a fall from a height or precipice, an attack of robbers or of quadrupeds. And thus, if Mars hold dominion, he will produce the misfortune by means of fire or wounds, through quarrels or by robbers; and if Saturn, it will be caused by a fall, by shipwreck, or by convulsive fits or spasms.

" The minor bodily disorders mostly occur on the Moon being posited in a tropical or equinoctial sign.

" Considerable diseases, however, take effect when the malefics may be configurated in the same situations as those before described, yet differing in one respect; that is to say, being occidental of the Sun and oriental of the Moon. In such cases, Saturn will generally produce spasmodic colic, phlegm, rheumatism, emaciation, sickliness, jaundice, dysentery, cough, obstruction, or scurvy; and, in women, besides these diseases, uterine disorders. Mars will cause expectoration of blood, rupture of bloodvessels, pulmonary attacks, atrabilarious attacks, ulcers, diseases of the generative and urinary organs, fistula, etc.; in women, to these calamities, he adds abortion, uterine hæmorrhage, excision of the fœtus or its mortification.

" Mercury, when acting with them, will contribute to the increase of the evil: thus, if he be in familiarity with Saturn he will greatly augment the coldness, and promote the continuance of rheumatism, and the disturbance of the fluids ; especially in the chest, throat, and stomach. If in familiarity with Mars, he will tend to procure greater dryness, and will render worse ulcers, abscesses, loss of hair, erysipelas, tetters, *insanity*, *epilepsy*, and similar disorders.

" Under the circumstances above detailed, the disease or hurt will be incurable,[2] provided there shall not be one of the benefics in configuration with the malefics which effect the evil, nor with the luminaries posited in angles ; and even though the benefics may be so configurated, the misfortune will still be incapable of remedy if the malefics be well fortified, and in elevation above them.

" Should the benefics, however, hold principal situations, and be in elevation above the obnoxious malefics, the disease or hurt will then be moderate, and have neither deformity nor disgrace attached to it ; and it will sometimes be altogether prevented and set aside, if the benefics be oriental. Jupiter, for instance, by means of human aid, such as wealth and rank can command, will conceal and soothe hurts and diseases ; and, if Mercury be joined with him, the assistance will be further improved by the addition of skilful doctors and good medicine. Venus, likewise, will ameliorate

---

[2] The improvement of the surgical art, at the present day, would render some of these cases curable.

diseases by medicine.   Lastly, should Saturn be present
in the configuration, the afflicted persons will move
abroad to show their maladies, and to complain; and, if
Mercury also be present, they will do so for the sake of
deriving support and profit from the exhibition."

In the " Text-Book of Astrology," vol. i., pp. 127-
135, the student will find many cases given in illustration of these aphorisms.   I have never yet found a single
case of privation of sight, nor of epilepsy, nor of insanity,
nor of any of the forms of consumption, in which the
horoscope was not in complete accordance with the rules
laid down by Ptolemy.   I have repeatedly challenged
those who condemn judicial astrology to put these rules
to the test by obtaining the moment of birth of persons
who were born blind or became so soon after birth, of
insane persons, and of epileptics, on condition that the
result, whether *pro* or *con*, shall be published to the
world.   The presence of the afflicting planet in the sign
Aries is said to cause the head and face to be chiefly
affected.   In *Taurus*, the throat.   In *Gemini*, the arms
and shoulders.   In *Cancer*, the chest, stomach, and
breasts.   In *Leo*, the heart and spine.   In *Virgo*, the
intestinal canal, and viscera of the abdomen.   In *Libra*,
the reins, loins, and bladder.   In *Scorpio*, the generative organs.   In *Sagittarius*, the hips and thighs, and
the os sacrum. In *Capricornus*, the knees. In *Aquarius*,
the legs and ankles.   In *Pisces*, the feet and toes.   In
fiery signs (Υ, Ω, and $\uparrow$), generally acute fevers; violent
accidents.   In *earthy* signs (δ, $\mathfrak{m}$, and $\mathcal{V}$), chronic
diseases, and consumption.   In *airy* signs (II, $\triangle$, and $\approx$)
diseases of the blood and skin.   In *watery* signs ($\mathfrak{S}$, $\mathfrak{m}$,
and $\mathcal{H}$), diseases of the lungs, and all that proceed from
taking cold; accidents by water.

Ptolemy says: " Pierce not with iron that part of the
body which may be governed by the sign actually

occupied by the Moon." This *may* be superstitious, but it would be advisable for operating surgeons to observe it when possible. An *Ephemeris* will show what sign the Moon occupies. I well remember a case of ovariotomy performed, by a very skilful surgeon, on the 13th of December, 1865, when the Moon was passing through *Scorpio* and applying to *conjunction* with Saturn. The poor woman died on the following day. Now, ovariotomy is an operation that can be delayed until a "favourable time." As Solomon said, there is "a time to kill and a time to heal." Ovariotomy is a very serious and dangerous operation, yet it is performed with wonderful success now-a-days. When possible, it would certainly be wise to choose for operations those days when the Moon is favourably configurated with Jupiter or Venus, and to eschew those days on which she is in evil aspect with Mars, Saturn, or Uranus. In hospitals it is usual to set apart one day in the week for major operations that admit of delay. This could be easily changed when evil configurations obtain. It would be better to run the risk of being deemed superstitious than to neglect an observance that might give the poor patient a better chance of recovery.

CHAPTER XV.

ON THE MIND AND DISPOSITION.

"Now since thou hast, although so very small,
Science of arts so glorious!"

HOMER'S HYMN TO MERCURY.

"CERTAINLY," says Bacon, "there is a consent between the body and the mind; and where Nature erreth in the one, she ventureth in the other. *Ubi peccat in uno, periclitatur in altero.*"

The planet Mercury has chief dominion over the mental faculties; the sentient, and the passions, are governed by the Moon and the ascendant.

Horace asks: "Why does one brother like to lounge in the forum, to play in the campus, and to anoint himself in the bath so well, that he would not put himself out of the way for all the wealth of the richest plantations of the East; while the other toils from sunrise to sunset for the purpose of increasing his fortune?" Horace attributes this diversity of character to the influence of genius and the *natal star.* The one is of the Venus type, the other of the Mercurial combined with the Saturnine.

The late Prince Consort was born with Mercury rising in the sign *Virgo*,[1] and this harmonises with his well-known character, his unwearied industry, his great talents, and his excellent judgment.

The progressive explorations of physiological and pathological science have demonstrated intimate bonds of union between mind and body unsuspected by any except astrologers.

---

[1] See his horoscope at p. 60, vol. i., of the "Text-Book of Astrology."

Robert Hooke, F.R.S., who was born on the 18th of July, 1635, at Freshwater, Isle of Wight was, like Newton, a sickly child. "His faults were," says a writer in the *Edinburgh Review* (July, 1880), "warpings of the mind, closely dependent, perhaps, on his unfortunate physical constitution. In spirit, as well as in person, Nature had set him somewhat awry. It was his misfortune that he could neither win sympathy nor inspire pity. His talents earned for him patronage; but his peculiarities repelled friendship. He lived sixty-eight years without attaching to himself a single human being, and died only to make room for his rival. Yet his intellectual qualities did not demand admiration more than his moral failings claimed tenderness. For surely infirmity has been rarely combined with genius in more painful and pitiable guise than in Robert Hooke." Here we have an instance of a Saturnine mind and disposition.

Writers on astrology give rules for describing, from the positions at birth, both the personal appearance and disposition. In regard to the former, they do not succeed so well as in the latter. When a planet is exactly rising at a birth, its character is strongly impressed on the child; and when such planet is closely configurated with the Moon and Mercury the disposition and mental characteristics are as strongly impressed as the physical. But when no planet happens to be rising, the task of describing the personal appearance (at maturity) is difficult and is not often successfully achieved.

In the case of the late Prince Consort, Mercury rising in Virgo, describes his personal appearance very closely; and, again, in the case of H.R.H. the Prince of Wales, Jupiter rising in Sagittary exactly describes his personal appearance.

Generally speaking, Ptolemy's aphorisms as to the influence of the several planets on the "form and temperament of the body," are pretty well borne out:—

"SATURN, when oriental [*i.e.*, within 5° of the ascendant], acts on the personal figure by producing a yellowish complexion and a good constitution ; with black or curled hair, a broad and stout chest, eyes of ordinary quality, and a proportionate size of body, the temperament of which is compounded principally of moisture and cold.

"JUPITER, oriental, makes the person white or fair, with a clear complexion, moderate growth of hair, and large eyes, and of good and dignified stature ; the temperature being chiefly of heat and moisture.

"MARS, ascending, gives a fair ruddiness to the person, with large size, a healthy constitution, blue or grey eyes, a sturdy figure, and a moderate growth of hair, with a temperament principally of heat and dryness.[2]

"VENUS operates in a manner similar to that of Jupiter, but at the same time more becomingly and more gracefully; producing qualities of a nature more applicable to women and female beauty, such as softness and greater delicacy. She also particularly makes the eyes beautiful, and renders them of an azure tint.

"MERCURY, when oriental, makes the personal figure of a yellowish complexion, and of stature proportionate and well-shaped, with small eyes and a moderate growth of hair ; and the bodily temperament is chiefly hot.

"The SUN and MOON, when configurated with any one of the planets, also co-operate : the Sun adds a greater nobleness to the figure, and increases the healthiness of the constitution ; and the Moon generally con-

---

[2] I have always found. when Mars is exactly rising at birth, that there is a mark or scar in the face, near the eyes.

tributes better proportion and greater delicacy of figure, and greater moisture of temperament."

In regard to mind and disposition:—

" SATURN, configurated with Mercury, makes men inquisitive, studious of law and of medicine, mystical, cunning, familiar with business, meditative, petulant, and fond of employment.

" JUPITER, configurated with Mercury, renders men fit for much business, fond of learning, and of geometry and the mathematics ; acute, temperate, well-disposed, skilful in counsel, beneficent, able in government, pious, valuable in all useful professions, ready in acquiring knowledge, philosophical, and dignified.

" MARS, connected with Mercury, renders men skilful in command, strenuous, active, obstinate, versatile, inventive, busy in all things, imposing, deceitful, inconstant, maliciously artful, inquisitive, fond of strife, and successful.

" VENUS, conciliated with Mercury, makes men lovers of the arts, philosophical, of scientific mind and good genius, poetical, delighting in learning and elegance, polite, voluptuous, luxurious in their habits of life, merry, friendly, fitted for various arts, intelligent, not misled by error, quick in learning, self-teaching, copious and agreeable in speech, serene and sincere in manner, delighting in exercise, honest, judicious, high-minded.

" MERCURY alone, having dominion of the mind, and being in a glorious position [i.e. in the ascendant], renders it prudent, clever, sensible, capable of great learning, inventive, expert, logical, studious of nature, speculative, of good genius, emulous, benevolent, skilful in argument, accurate in conjecture, adapted to sciences and mysteries, and tractable. But when placed contrarily [i.e. weak and afflicted, and without the support of either of the benefics], he makes men busy in all things, precipitate, forgetful, impetuous, frivolous, variable, regretful, foolish, inconsiderate, void of truth, careless, inconstant, insatiable, avaricious, unjust, and altogether of slippery intellect, and predisposed to error.

"To these influences and their effects, as above detailed, the MOON also contributes; for, should she be in extreme latitude, she will render the properties of the mind more various, more versatile in art, and more susceptible of change: if she be in her node, she will make them more acute, more practical, and more active. Also, when in the ascendant, and during the increase of her illumination, she augments their ingenuity, perspicuity, firmness and expansion; but, when found in her decrease, or in occultation, she renders them heavier, more obtuse, more variable of purpose, more timid, and more obscure.

"The SUN likewise co-operates, when conciliated with the lord of the mental temperament; contributing, if angularly posited, to increase probity, industry, honour, and all laudable qualities; but, if adversely situated, he increases debasement, depravity, obscurity, cruelty, obstinacy, moroseness, and all other evil qualities."

Down to the publication of the "Text-Book of Astrology" (in January, 1880), it was the fashion of writers on astrology to assert that the Sun in conjunction with Mercury, at birth, "destroys the mental abilities of the native."

This is a fallacy, as I have shown elsewhere.

The question of insanity was very ably handled by Mr. A. G. Trent in his treatise on "The Soul and the Stars," which appeared in the *University Magazine* for March, 1880, and was mainly reprinted in *Urania* for April and May, 1880.

"Nothing can be simpler," says Mr. Trent, "than the

rules respecting insanity which have come down to us from Egyptian and Chaldæan antiquity. It is, that mental disease is liable to occur when Mars and Saturn (to which modern research has added Uranus), are at birth in *conjunction, quadrature,* or *opposition* with Mercury and the Moon, but Mercury more particularly. It is by no means asserted that insanity always, or even often, occurs with such positions; what is asserted is, that it rarely occurs without it. When controlled by favourable influences it may even be beneficial, on the principle that a spice of the devil is a desirable ingredient in the composition of a good man. When no such influences exist the most ordinary result is moral obliquity, a practical demonstration of the profound truth that wickedness *is* madness.

"We begin by instancing nine sovereign princes, notoriously insane or deficient in intellect, upon whose birthdays Mercury or the Moon, or both, will be found to have been affected by Mars, Saturn, or Uranus, in the manner described. They are: Paul, Emperor of Russia; George III., King of England; Gustavus IV., King of Sweden; Ferdinand II., Emperor of Austria; Maria, Queen of Portugal; Charlotte, Empress of Mexico; Charles II., King of Spain; Murad V., Sultan of Turkey; and Constantine of Russia (who abdicated in favour of his brother). The planetary positions, so far as essential for our present purpose, are as follow:—

Emperor Paul.	George III.	Gustavus IV.
Oct. 1, 1754.	June 4, 1738.	Nov. 1, 1778.
☿ 5° ♎	☿ 25° ♊	☽ 22° ♓
☽ 10° ♈	☽ 10° ♑	♅ 19° ♊
♄ 15° ♑	♅ 5° ♑	♂ 19° ♍
	♄ 27° ♊	
	♂ 7° ♈	

Emp. Ferdinand.	Qu. Portugal.	Emp. Charlotte.
April 19, 1793.	Dec. 17, 1734.	June 7, 1840.
☿ 9° ♉	☿ 17° ♐	☿ 11° ♋
☽ 16° ♌	♅ 19° ♐	☽ 16° ♍
♅ 19° ♌	♂ 19° ♓	♅ 20° ♓
♄ 4° ♉		♄ 18° ♐
		♂ 8° ♋

Chas. II. of Spain.	Murad V.	Constantine.
Nov. 6, 1661.	Sept. 21, 1840.	May 8, 1779.
☿ 26° ♏	☿ 24° ♍	☿ 24° ♉
♄ 25° ♏	☽ 4° ♌	☽ 15° ♒
	♅ 18° ♓	♄ 24° ♏
	♄ 16° ♐	♂ 22° ♏

"Is this chance? most people, perhaps, will at first regard this as the lesser improbability. We therefore follow up the inquiry by adducing six insane persons of genius. Gèrard de Nerval, who committed suicide in a fit of insanity; Alfred Rethel, the painter of Der Tod als Freund; Agnes Bury, the actress; Jullien; Pugin; and Paul Morphy:—

G. de Nerval.	Rethel.	Agnes Bury.
May 21, 1808.	May 15. 1816.	April 28, 1831.
☿ 22° ♉	☿ 6° ♊	☿ 27° ♉
♄ 18° ♏	☽ 13° ♑	☽ 23° ♏
♂ 22° ♉	♅ 10° ♐	♄ 24° ♌
	♂ 13° ♋	

Jullien.	Pugin.	Morphy.
April 23, 1812.	March 1, 1812.	June 22, 1837.
☿ 21° ♉	☿ 22° ♎	☿ 12° ♊
☽ 23° ♍	☽ 16° ♎	♅ 8° ♓
♅ 22° ♏	♅ 22° ♏	♂ 9° ♍
♄ 8° ♑	♂ 24° ♈	

" We next take four instances of highly gifted men who lost their faculties in old age :—

Swift.	Southey.
Nov. 30, 1667.	Aug. 12, 1774.
☿ 9° ♑	☿ 0° ♍
☽ 11° ♎	☽ 14° ♎
♂ 8° ♎	♅ 2° ♊
Moore.	Faraday.
May 28, 1779.	Sept. 22, 1791.
☿ 17° ♉	☿ 19° ♎
☽ 17° ♏	☽ 22° ♋
♄ 22° ♏	♄ 16° ♈
♂ 16° ♏	

" Compare with these the cases of three mischievous lunatics, the would-be assassins of the late and present kings of Prussia, and a remarkable case of a female lunatic described in the *Revue des Deux Mondes* for January 15th, 1880:—

Sefeloge.	Nobiling.
March 29, 1821.	April 10, 1818.
☿ 1° ♈	☿ 23° ♓
☽ 11° ♒	☽ 19° ♋
♅ 0° ♑	♅ 18° ♈
	♄ 19° ♓
	♂ 25° ♊
Oscar Becker.	G———.
June 18, 1839.	Jan. 2, 1843.
☿ 16° ♊	☿ 14° ♑
☽ 21° ♍	☽ 1° ♒
♅ 11° ♓	♄ 16° ♑
♂ 24° ♍	♂ 2° ♏

" Sefeloge has Mercury in quadrature with Uranus, and the Moon in semi-quartile with both; Nobiling, Mercury in opposition with Saturn and quartile with

Mars, the Moon in quartile with Uranus[3]; Becker, Mercury in quartile with Uranus, Moon, and Mars, and the two latter in opposition with the former. The French lunatic has Mercury in conjunction with Saturn, and the Moon in quartile with Mars.

" To recapitulate, we think it has been shown that quartile and opposition aspects between Mercury and the Moon on the one hand, and Mars, Saturn, and Uranus on the other, will be found co-existent either with insanity or with the quick, restless, and imaginative temperament most liable to mental disturbance. This general proposition is of course liable to the most extensive modifications according to the strength of these planets at the time of birth, and to the influence of the benefic planets. It holds equally true of the affections of the Sun, Moon, and degree ascending as respects the physical constitution ; and of the Sun, Moon, and meridian as regards success in life. We do not deny the existence of many difficulties and anomalies, and fully admit that astral science is incompetent to explain the divergences of human constitution and character without a free use of the doctrine of heredity. Our contention is that the two theories complete each other, the latter accounting for the element of stability, the former for the element of variability.

" It will be conceded that there is nothing occult or mystical in the line of argument we have been pursuing. We have appealed throughout to the testimony of facts of history and biography, partly astronomical observations derived from no more recondite source than the ordinary ephemeris. Any one can verify or disprove these observations in a moment by the same process;

---

[3] At Nobiling's birth Mars was on the place of Saturn at the Emperor's birth.

any one who will be at the trouble to search for examples can investigate the subject for himself."

As Dr. Sharp says:[4] " The diseases of the body act through the vital principle upon the mind, and, on the other hand, the disorders of the mind act through the same medium upon the body." When this occurs in a marked degree, it is found that both the Moon and Mercury are afflicted by the malefic planets. The term *lunacy* is derived from the observations of the ancients that the Moon was usually afflicted at the birth of persons who became insane. A remarkable proof of the influence exerted by the planets on the mental faculties is afforded in the case of monomania in twins recorded in the *Psychologie Morbide* by Dr. J. Moreau (de Tours), Médecin de l'Hospice de Bicêtre; and another, in the case of insanity in twins described by Dr. Baume, in the *Annales Médico-Psychologiques*.[5] Born under the same celestial influences, when mania attacked one of the twins the other quickly fell a victim to it also.

In order to predicate *mens sana in corpore sano*, it is necessary to find in the horoscope that Mercury, the Moon, the Sun, and the ascendant, are free from affliction, and supported by the benefic planets.

---

[4] " Essays on Medicine."
[5] 4 Série, vol. i., 1863, p. 312.

CHAPTER XVI.

ON DESTINY.

"Follow but thy star,
Thou can'st not miss at last a glorious haven ;
Unless in fairer days my judgment erred,
And if my fate so early had not chanc'd
Seeing the heavens thus bounteous to thee, I
Had gladly given thee comfort in thy work."

DANTE.

THE belief in judicial astrology is *not* fatalistic. Astrologers have ever taught that the planets *agunt, non cogunt;* they act, or incline, but nowise compel.

Man has free will to choose his path in life, but he cannot control his destiny entirely. He may speak of himself as "the architect of his own fortune"—when he happens to amass wealth. But he who falls into comparative or actual poverty complains of his "luck." A man may "take up with avarice" as "a gentlemanly vice," and he will be greatly inclined to do so if Saturn be his ruling planet—he may make wealth his god ; or, he may despise riches, and leave it to what he calls "chance" whether they come to him or not, directing his talents to the amelioration of the lot of his unfortunate fellow-creatures, or to the improvement of science. He may be a man of pleasure, or he may so over-work himself as to induce early death. He may be a religious man or a "free-thinker."

The wise man of old observed that "the race is not always to the swift, nor the battle to the strong; neither

yet bread to the wise, nor yet riches to men of understanding; nor yet favour to men of skill ; but time and
chance [or, literally, *time of good events and of evil
events*] happeneth to them all."

Talent, indomitable energy, pluck, and perseverance,
combined, and favoured with good health, will achieve
wonders, and will enable a man born in an evil hour to
make his mark in the world. Nevertheless, a man
cannot be termed fortunate who has to fight an uphill
battle against tremendous odds. Only the lucky few
find their course of life run smooth.

Astrology shows who are born fortunate and who are
born unfortunate. The use of it is to point out the
career and the part of the world in which success will be
most readily gained.

Many who have been born in the purple have come
to grief, some through no fault of their own. Others,
like Napoleon I., have fought their way from the ranks
to the highest pinnacle of power. Napoleon I. was so
unfortunate as to fall from his proud position: possibly
had Jupiter instead of Saturn been in the midheaven at
his birth, he might have died on the throne.

To succeed in life, easily, and to retain what is won,
it is indispensable that the Sun, Moon, ascendant and
meridian be free from the affliction of the malefics, and
be supported by the benefics. Jupiter located in the
ascendant, second house, or meridian, receiving the
application of the Moon, or configurated with the Sun,
is very fortunate. Saturn so situated, usually brings
ruin or severe misfortune (if not early death) in the
end.

The nativities of the Bourbon family afford a most
striking illustration of this fact. The following letter
was printed in the *Spectator*, February 22nd, 1862:—

"The economy of your journal does not permit me to

H

print the horoscopes of the Bourbons in full, but I give the exact times of birth, so that anyone who will bestow a few hours' attention on the elementary principles of astrology may be enabled to draw them for himself. With the exception of that of Louis Philippe, for which I do not vouch, these times of birth are all derived from official documents which anyone who pleases may inspect at the British Museum.

"Louis XVI., born August 23rd, 1754, at 6ʰ 24ᵐ a.m. Mars rising, Uranus setting; Saturn in sesquiquadrate aspect with the Sun, also afflicting the Moon, and the latter, thus rendered malefic, in square aspect with the Sun, again.

"Marie Antoinette, November 2nd, 1775, 7ʰ 30ᵐ p.m. Uranus approaching the meridian in sesquiquadrate with the Sun; the Moon exactly between Mars and Saturn.

"Louis XVII., March 7th, 1785, 7ʰ p.m. Uranus in square with the Sun.

"Princess Elizabeth (guillotined), born May 3rd, 1764, 2ʰ a.m. Mars culminating, Saturn in conjunction with the Sun.

"Louis XVIII., born November 17th, 1755, 4ʰ a.m. Mars in conjunction with the meridian, Saturn in opposition to both, but *Jupiter rising;* therefore, after all his vicissitudes of fortune, he died upon the throne.

"Charles X., born October 9th, 1757, 7ʰ p.m. Saturn and Uranus in conjunction near the meridian, in square to Jupiter, Mars in opposition to the meridian; no favourable indication of any kind.

"Duc de Bordeaux, September 29th, 1820, 2ʰ 35ᵐ a.m. Saturn in opposition with the Sun.

"Duchess of Parma (his sister), September 21st, 1819, 6ʰ 35ᵐ a.m. Mars in conjunction with the meridian, Saturn in opposition with the Sun.

"Duke of Angoulême, August 6th, 1775, 3ʰ 45ᵐ p.m. Mars and Saturn in conjunction with the meridian, in square with Uranus, and all three in semisquare with the Sun.

"Duchess of Angoulême, December 19th, 1778, 11ʰ 25ᵐ a.m. Uranus in opposition to both the Sun and the meridian.

"Duc de Berri (assassinated), born January 24th, 1778, 11ʰ 15ᵐ a.m. The Sun in square and the Moon in opposition with Uranus.

"Louis Philippe, October 6th, 1773, 9ʰ 40ᵐ a.m. Saturn culminating, afflicting the Moon, but some indications of good fortune.

"Duc de Nemours, October 25th, 1814, 5ʰ p.m. Saturn culminating.

"Prince de Joinville, August 14th, 1818, 1ʰ 40ᵐ p.m. Mars in conjunction with the meridian, Saturn in opposition to both.

"Duc d'Aumale, January 14th, 1822, 9ʰ p.m. The Sun in square with Saturn, in sesquiquadrate with Mars.

"Duchesse d'Aumale, April 26th, 1822, 6ʰ 15ᵐ p.m. Mars culminating in sesquiquadrate with Uranus.

"Duc de Montpensier, July 31st, 1824, 5ʰ 40ᵐ p.m. Saturn culminating, but Jupiter in good aspect with the meridian ; and the fact is that the duke, having become a Spanish prince by marriage, has suffered comparatively little by the late revolution.

"I have not been able to procure the times of birth of the Duke of Orleans and the Queen of the Belgians, but the Princess Clémentine of Gotha (June 3rd, 1817, 1ʰ 40ᵐ a.m.) has not one indication of evil in her horoscope, and she has, accordingly, *been entirely exempt from the misfortunes of her family.* In the nativity of the Comte de Paris, on the other hand (August 24th,

1838, 2^h 45^m p.m.), the Sun is afflicted by all the malefics at once. I have only to add that the indications of character in these various horoscopes are, to the best of my judgment, quite as accurate as the indications of destiny."

The nativities of Napoleon I. and Louis Napoleon III. show Saturn in the meridian, and afflicting the Moon. The former is given at page 210 of the "Text Book of Astrology," vol. i. Here is the horoscope of Louis Napoleon:—

FIG. 5.

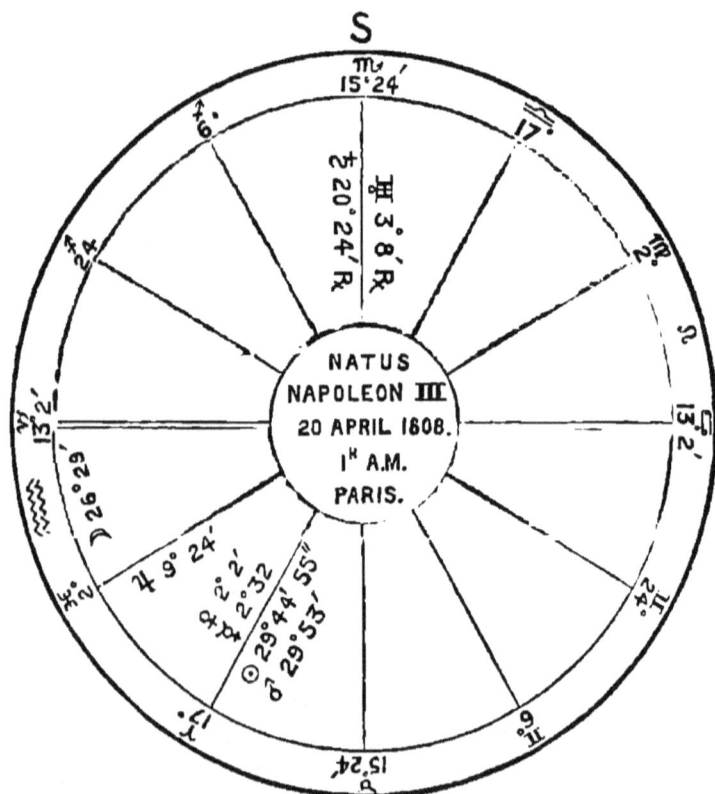

SPECULUM.

	Lat.		Dec.		R.A.		M.D.		S.-Arc.	
☉	*	*	11	24 N	27	40	15	16	76	40
☽	5	14 N	7	46 s	326	58	75	58	98	58
☿	2	31 s	1	18 s	3	19	39	37	91	29
♀	1	34 s	0	38 s	2	28	40	28	90	43
♂	0	21 s	11	7 N	27	55	15	1	77	1
♃	0	56 s	8	55 s	341	22	61	34	100	20
♄	2	29 N	15	23 s	228	38	5	42	71	33
♅	0	31 N	12	3 s	211	4	11	52	75	52

The R.A. of the meridian is $14^h 51^m 44^s$, or $222^\circ 56'$ in arc.

In the horoscope of Napoleon I., Saturn was in the midheaven and in opposition to the Moon; in that of Napoleon III., Saturn was in the midheaven and in square aspect with the Moon. Both Emperors fell from power and died in exile; both at certain periods of their lives were made prisoners.

The symbols of a *fortunate* and *successful* career are: 1. The Sun and Moon in mutual benefic aspect (the *sextile* or *trine*), one of them being in the meridian. 2. The Sun or Moon applying to the *conjunction*, *sextile*, or *trine* aspect with Jupiter, one of them being angularly posited. The square and opposition of Jupiter with the Sun or Moon generally bring success in the end (unless Jupiter be afflicted by the malefics), but it is usually attended with delay, heavy expenses, and some difficulties. 3. The Sun attended by one fortune and the Moon by the other, one or both of the luminaries being in an angle. [This was the case at the birth of Goethe.] 4. The Moon applying to the conjunction with Jupiter in the second house, and free from affliction.

The signs of an *unfortunate* career are: 1. The Sun, Moon, ascendant and midheaven afflicted by Saturn and unsupported by either of the fortunes. 2. The Sun or Moon applying to the *conjunction, square,* or *opposition* with Saturn, and neither luminary configurated with the fortunes.

Jupiter angular and in conjunction, sextile or trine aspect with either the Sun or Moon, augurs success in the Senate, at the bar, or by means of University or Church appointments. Also in trade generally.

Venus angular and in conjunction or par. dec. with the Sun, or in conjunction, sextile, or trine aspect with the Moon, gives wealth through the friendship or kind offices of ladies of position and influence, by marriage, or by trading in articles of commerce ruled by Venus.

Mars happily configurated with the Sun or Moon augurs success in the army, the practice of surgery, and in any mechanical trade in which iron, sharp tools, or fire may be used.

Saturn happily configurated with either the Sun or Moon is favourable for the pursuit of agriculture, building, dealing in land or houses, mines, etc.

Mercury angular and either happily configurated with the Moon, or in conjunction or par. dec. with the Sun, and free from affliction, augurs success in the pursuit of science, literature, art, merchandise, or travelling.

Persons in whose horoscopes the ascendant, midheaven, the Sun or Moon, is afflicted by Saturn should never speculate, and should be extremely cautious in dealing in land, houses, mines, etc. Persons in whose horoscopes either the ascendant, midheaven, the Sun or Moon, is afflicted by the square or opposition aspect of Mars, should avoid danger as much as possible, and should eschew martial pursuits and quarrels.

When the fortunate planets or luminaries are in the eastern part of the heavens at birth, the person then born will do best in a *direction* east of his birthplace; and, *mutatis mutandis*, the same remark applies to the other parts of the heavens.

The *signs* containing the Sun, Moon, and planets will generally indicate the parts of the world most favourable for health and affairs. For example: A gentleman who was born when the Moon was in the sign *Cancer* and applying to the conjunction with Jupiter (in the second house) went to a seaside place (♋ is a sign of the "watery" triplicity), and made a fortune by shipping.

For *fame*, take the sign containing the Sun or Moon (whichever may be angular, unafflicted, or supported by a benefic) as the index.

For *business, science,* or *literature*, take the sign containing Mercury (if free from affliction); otherwise that containing Mars. For "*luck*," if every other means fail, try the sign containing Venus or Jupiter—whichever may be the stronger.

Avoid those places ruled by the signs containing the planet or planets afflicting either the luminaries, the ascendant or midheaven. A gentleman who was born when ♂, ♄, and ♅ were together in *Aries* and afflicting the Moon, went (being ignorant of Astrology) to Birmingham (ruled by ♈), to seek his fortune. He lost everything he possessed there, and contracted small-pox (a martial disease) into the bargain.

To command success in the law, Mercury and Jupiter must be strongly placed in the horoscope and favourably configurated with either the Sun or Moon.

To insure success in the pursuit of the medical profession, or in the army, Mars should be strongly posited and happily configurated with either the Sun, Moon.

Mercury, or the midheaven. In like manner, judge of probable success in other pursuits by the position and configuration of the ruling planet.

In *Urania* for May, 1880, there is an interesting paper on " Successful Commanders," in illustration of the fact that at the birth of illustrious soldiers either the Sun, Moon, or Mercury is generally found favourably configurated with Mars. The most constant configuration was the *trine* of the Moon with Mars; next the *sextile* and *opposition*, and the *square* aspect was seldom met with. The Duke of Wellington was born when the Sun was in sextile with Mars, the Moon in sextile with the Sun and in trine with Mars, while Mercury was in the martial sign *Aries*. Moltke has the Sun and Mercury in *Scorpio* (the most martial of all the signs), the Moon in square and Mercury in opposition to Mars. Lord Nelson had Mars rising in *Scorpio*, and in sextile with the Moon. Sir Frederic Roberts, the hero of Candahar, was born on the 30th of September, 1832, in India (hour unknown); the Sun was in trine with Mars.

The late Lord Chief Justice, Sir Alexander Cockburn, was born on the 24th of December, 1802; the Sun (at noon) was in ♑ 2° 2′ 47″, the Moon in ♐ 26° 51′, Mercury in ♐ 23°, and Jupiter in ♎ 4° 47′. Hence the Sun was in square aspect with Jupiter, and the Moon was in conjunction with Mercury—symbols of industry, talent, and success. Mark Firth, who made a large fortune and became a public benefactor, was born on the 25th of April, 1819, when the Moon was in conjunction with Mercury and in square aspect with Jupiter. Sir Humphry Davy was born on the 17th of December, 1778, when the Sun was in square aspect with Jupiter.

FIG. 6.

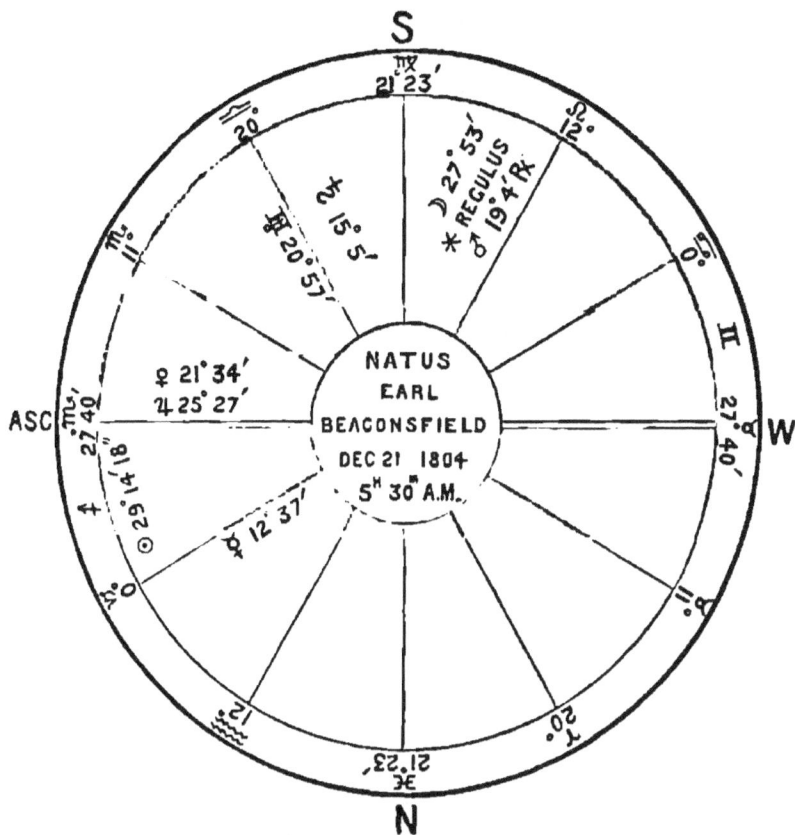

SPECULUM.

	Lat.		Dec.		R.A.		M.D.		S.-Arc.	
☉	✳	✳	23	28 s	269	10	82	55	123	5
☽	2	47 s	9	36 N	149	6	22	59	102	17
☿	2	12 s	25	4 s	283	57	68	8	126	2
♀	2	3 N	16	3 s	229	41	57	36	68	47
♂	3	19 N	18	15 N	142	35	29	30	114	30
♃	0	53 N	18	18 s	233	19	61	14	65	25
♄	2	23 N	3	45 s	194	48	22	43	85	16
♅	0	37 N	7	37 s	199	34	27	29	80	19

The R.A. of the midheaven is 11^h 28^m 19^s or
$172°$ $4'$ $50''$.

The horoscope of the late Prime Minister, Earl Beaconsfield, shows Jupiter and Venus rising, the Moon having the square aspect with Jupiter and the trine with the Sun; Mercury in the second house. His indomitable energy is indicated by the proximity of the Moon to Mars and the martial star Regulus; his patience, tenacity of purpose, and diplomatic skill, by the connexion of Mercury with Saturn; his amiable disposition and good fortune, by the ascending position of Jupiter and Venus and their connexion with the Moon ; his literary talent by the configurations of Mercury with the ascending degree, the Moon, and Mars.

With this brilliant horoscope—so pre-eminently symbolical of the genius and splendid achievements of the late Earl, who had to contend with so many difficulties and so much misrepresentation, and yet came off victorious—we will bring these remarks on destiny to a close.

The importance of martial *traits* in the formation of the character of heroes and heroines, in all the various grades of life, is recognised by Dante, in his *Paradiso*, Canto XVII. Writing of the heaven of the planet Mars, he says:—

> "With him shalt see
> That mortal, who was at his birth imprest
> So strongly with this star, that of his deeds
> The nations shall take note."

CHAPTER XVII.

ON MARRIAGE.

" Whose souls do bear an equal yoke of love,
There must be needs a like proportion
Of lineaments, of manners, and of spirit."

SHAKESPEARE.

SYMPATHY is quite as essential as mutual love to insure happiness in marriage. Shakespeare understood and insisted on this. In friendship, as in love, sympathy is chiefly produced by the harmonious configurations of the Sun and Moon in the respective nativities. Liking and disliking, loving and hating, are thereby explained. In love, the concurrence of Venus and Mars is an additional and a most powerful testimony. " Love at first sight" is often due to such concurrence and to the harmonious configuration of the luminaries. The same sign ascending at the moments of birth of two individuals will, if they happen to meet, produce great friendship. When the ascendant of one is the midheaven of the other, friendship results, and the one whose ascending sign is in the midheaven of the other usually becomes a benefactor to such person. If the benefic planets (Jupiter and Venus) be configurated with the luminaries in both nativities (of friends or lovers), the friendship will prove advantageous and pleasurable to both persons. If, on the other hand, the malefics afflict the luminaries, the " course of true love" will not " run smooth," and, like Romeo and Juliet, it will be a case of " a pair of star-crossed lovers."

The ancients observing this, made a serious attempt

to prevent the manifold evils which too often result
from persous being " unequally yoked." This is mani-
fest in the following aphorisms of Claudius Ptolemy, the
astrologer:—

" Whenever both nativities, viz., that of the husband
and that of the wife, may exhibit the luminaries configurated together in concord, that is to say, either in
trine or in *sextile* to each other, the cohabitation will
most usually be lasting; especially if the said concord
exist by means of an interchange [or by " mutual reception "]; but its duration will be also much more
securely established provided the Moon in the husband's
nativity should correspond or agree with the Sun in the
wife's nativity. If, however, the relative positions of
the luminaries be in signs inconjunct,[1] or in opposition,
or in quartile,[2] the cohabitation will be speedily dis-
solved upon slight causes, and the total separation of
the parties will ensue.

" And should the configuration of the luminaries,
when made in concord, be aspected by the benefics, the
cohabitation will continue in respectability, comfort,
and advantage; but, on the other hand, it will abound
in strife, contention, and misfortune, if the malefics be
in aspect to the said configuration.

" In like manner, even though the luminaries may
not be favourably configurated in concord, should the

---

[1] Signs considered by Ptolemy to be "inconjunct" are
those separated by a space of 150°, as are ♈ and ♍, ♉ and ♎,
♊ and ♏, ♋ and ♐, ♌ and ♑. It is an error, for the angle
of 150° is an "aspect."

[2] The quartile is not an impediment when the luminaries
are, in both nativities, configurated with the benefics and
free from the rays of the malefics, and when the aspects of
Mercury are similar.

benefics still offer testimony to them, the cohabitation
will then not be entirely broken off, nor totally destroyed
for life, but will be again renewed and re-established as
before. But if, on the contrary, the malefics bear
testimony to such discordant disposition of the luminaries, a dissolution of the cohabitation will take place,
accompanied by scorn and injury. Should Mercury
alone be conjoined with the malefics, it will be effected
by means of some public inculpation; and if Venus
also be found with them, it will be on the ground of
adultery," etc.

The moral of this is that when courtship has arrived
at the "Ask Papa" stage, an examination and comparison should be made by the parents or guardians, of
the respective nativities of the lovers, and the betrothal
should not be allowed if the horoscopes are found to be
discordant.

The study of astrology throws a flood of light on the
important question of matrimonial felicity or infelicity.
" Incompatibility of temperament" is often alleged to be
the cause of separation of married persons. By means
of this science such incompatibility could and should be
discovered before the knot is tied.

The late King of Denmark was very unhappy in
marriage and divorced two queens in succession, but
having become attached to a milliner he ennobled her,
espoused her morganatically, and lived very happily
with her until his death. *The Sun in the one nativity
was in the place of the Moon in the other.*

Goethe had the Sun in 5° 9' of ♍, Frau von Stein
had Mars in 5° of ♍, whereas Christiane Vulpius had
the Moon in 4° of ♍. The former was loved and for-
saken, the latter was loved and married by Goethe.
"This does not look like mere coincidence," says Mr.
Trent, in the *University Magazine* (March, 1880).

The same phenomena are repeated in the case of Novalis and the girl of thirteen for whom he conceived the intense attachment that has so puzzled his biographers. Sophie's Moon is in the same place as Novalis's Sun and Moon, and her Venus is in the place of his Mars.

The planet to which the Moon *applies* at birth, either by conjunction, sextile, or trine, generally effects marriage (by "direction"). The planet to which the Sun applies, in the horoscope of one of the fair sex, describes the nature of her husband. The planet to which the Moon applies, in the horoscope of a man, pre-signifies the nature of his wife. When the Sun, or Moon, is in a bi-corporal sign (♊, ♓, and the first half of ♐), or configurated with several planets, more than one marriage usually takes place. When Venus is in a bi-corporal sign and in aspect with Mars, the person then born has many love-engagements.

Mars configurated with the Sun or Moon, as the case may be, pre-signifies an irascible and rather intractable partner in marriage. Venus, so configurated, pre-signifies a handsome person, and one of an affectionate disposition. Mercury, one who is provident and clever. Jupiter, one who is benevolent, jovial, and honourable. Saturn, one who is steady, laborious, and constant.

Uranus when in conjunction, square, or opposition with the Sun at the birth of a woman, or with the Moon at birth of a man, causes much unhappiness in marriage; and when at the same time Uranus is evilly configurated with Venus, separation usually ensues.

Saturn in square or opposition with the Sun renders women very unfortunate in marriage. If either the Sun or Moon be in the seventh house and applying to conjunction or evil aspect with Saturn, the early death of the partner in marriage is pre-signified.

Women who have the Sun and Venus afflicted by Saturn or Uranus, the Sun having no configuration with either Mars or Jupiter, seldom marry.

Men who have the Moon and Venus so configurated, do not usually marry.

When Venus is much afflicted, whatever may be the configurations of the luminaries, the person then born is unfortunate in love and matrimony.

Shakespeare's marriage is said to have been unhappy, the usual fate of the poet, who rarely has the good fortune to choose as a mate one whose nativity sympathises with his own. At the marriage of Romeo and Juliet, Shakespeare expresses the desire for the good influences of the benefic planets, thus:—

"So smile the heavens upon this holy act,
That after hours with sorrow chide us not."

CHAPTER XVIII.

ON DIRECTIONS.

PRIMARY DIRECTIONS are arithmetical computations of the apparent motion of any point in the heavens, or of any heavenly body, from the situation it occupied at the moment of birth until it meets with the conjunction, parallel declination, or aspect of some other body or point. The value thus obtained is termed the " *arc of direction*," and it is converted into time by allowing every *degree* of arc to represent *one year* of life, and every *five minutes* over and above the number of degrees to represent *one month*.

All directions of the midheaven are measured by an arc of *right ascension*. Those of the ascendant *in mundo* by the semi-arc of the body directed to it, and in the *zodiac* by *oblique ascension*.

All directions of the Sun, Moon, and planets are computed by means of their semi-arcs. Those directions, which for the sake of classification are termed zodiacal, are really mundane. Primary directions are formed by the revolution of the earth on its axis, and all those that require to be computed for the ordinary term of life are formed within a few hours of the time of birth.

The place of a heavenly body in a nativity is considered as the body itself, as the heavenly bodies are believed to impress their several natures on the places held by them at the moment of birth as fully as if they were always located therein, although they may be no longer there when the "significator" arrives; thus, if we direct the Sun to the conjunction with Jupiter, we mean to the radical place of Jupiter.

Before commencing the computation of arcs of direction it is necessary to form a "speculum," such as that attached to the horoscope of Earl Beaconsfield (see page 106), containing the latitudes, declinations, right ascensions, meridian distances, and semi-arcs of the Sun, Moon, and planets.

The rules for computing primary directions are given in my "Text-Book of Astrology," vol. i. Numerous examples are also given. The effects ascribed to the various directions are stated therein, hence there is no necessity to reproduce them in these pages.

It will be well, however, to give a few additional instances of the nature of the events corresponding with that of the group of directions coinciding in the nativities of some eminent personages.

The horoscope of Her Majesty Queen Victoria is given at page 70.

The following directions and remarkable events agree with the principles of astrology. It will be observed that no rectification is made of the time of birth given in the official bulletin. There is no doubt that the time was very correctly given, and those astrologers who rectified it by bringing M.C. ♂ ♃ up to the period of the Queen's accession to the throne made a great mistake.

Her Majesty acceded to the throne on the death of her uncle, King William IV., on the 20th of June, 1837. The arc for this event is 18° 4'. The following train of directions then operated:—

		°	′
☉ ☍ et par. ♅ m., d.	....	18	0
☽ 144° M.C. zod., d.	....	18	4
☉ ☍ ♅ zod., d.	....	18	16
☽ ✶ ♀ zod., d.	....	18	17

On the 10th of February, 1840, the Queen was married (arc = 20° 45'), under the following directions :—

	° ′
☉ ☌ et par. ♀ m., con.	20 42
☉ ⚹ ♀ zod., d..........................	20 57

In May, 1850, Her Majesty was struck on the head with a stick by Pate. The danger then incurred is shown by the affliction of the Sun by Mars:—

	° ′
♂ par. ☉ zod., d.	31 8
☉ par. ♂ zod., con.	31 37

In Her Majesty's 36th year war was declared by England and France against Russia. In September, 1854, the battle of the Alma was fought, that of Balaclava in October, and Inkermann on the 5th of November. The fall of Sevastopol took place in September, 1855, and peace was dictated to Russia by the allies in the spring of 1856. These momentous events happened under the following directions:—

	° ′
Asc. ☌ ♄ zod., con.......	35 10
M.C. par. ♀ zod., d.	35 17
☉ ⚹ ☿ zod., d........................	35 29
M.C. ⚹ ☿ zod., d......................	35 53
M.C. □ ♄ zod., con.	36 0
☉ s. ⚹ ☽ m., d.	36 4
M C. ss. □ ☽ m., con.	36 10
M.C. △ ☿ m , con.	36 29
M.C. ⚹ ☉ m., d.	36 32

In the early summer of 1857 the Indian mutiny broke out, and, to the astonishment of the world, the British forces entirely suppressed it within eighteen months. This great crisis took place under the operation of

		°	′
☉ par. ♄ zod., con		37	57
M.C. par. ♅ zod., con		37	56
☉ △ ♃ m., d.		39	22
Asc. ✳ ♂ m., d.		39	34
☉ ✳ ☽ zod., con		39	50

The last three of the foregoing directions measure exactly to the period of the final suppression of the mutiny and of the Indian Empire being annexed to the Crown.

The Duchess of Kent died on the 16th of March, 1861. The arc for this event is 41° 9′, which is very nearly the same as the meridian distance of Uranus:—

		°	′
M.C. ♂ ♅ zod., con		41	56
M.C. ♂ ♅ m., con.		41	57

On the 14th of December, 1861, Her Majesty and the entire nation were plunged into the deepest grief by the death of the Prince Consort. This sad event took place under the operation of the following train of evil directions:—

		°	′
Asc. 45° ♀ zod., con		42	27
♄ ♂ ☉ zod., d.		42	43
Asc. 45° ☉ zod., d.		42	44
☽ 45° ☉ zod., d.		42	51

The war in Afghanistan, and the brilliant victories of General Sir Frederic Roberts on the heights before Cabul, occurred under the following directional influences:—

		c	′
☉ par. ☉ zod., d.		59	57
M.C. ♂ Antares, m., con		60	5
M.C. ✳ ♃ m., d.		60	29

The Zulu war, the disaster of Isandhlwana, and the victory of Ulundi, in 1879, took place under the above directions.

In the early part of 1880 the first symptoms were manifested of the disturbance in Ireland, which culminated in the reign of terror in that unhappy country. In April, 1880, Lord Beaconsfield's Government were defeated at the poll. These events happened under the following directions:—

	°	′
M.C. par. ♄ zod., d.	60	55
Asc. ⚹ ♅ m., c.	60	56
☽ □ ♄ m., d.	61	0

In August, 1880, the disaster to General Burrows' division took place in Afghanistan; and in the following month Sir Frederic Roberts marched to the relief of Candahar and defeated Ayoub Khan. On the 16th of December, 1880, the Boers raised the standard of revolt, and in a few weeks thrice defeated the British troops sent against them. These events coincided with the operation of the following directions:—

	°	′
⊙ △ ♄ zod., d.	61	11
♄ par. ⊙ zod., con.	61	36

The student will find that the nature of the directions enumerated corresponds exactly to the nature of the events happening during the period of their operation. Similar correspondence could be shown in the nativities of Napoleon I., Napoleon III., the Emperor of Germany, the late Czar of Russia, and in all the nativities of eminent personages whose times of birth have been accurately noted and recorded in the press, did space permit. It is open to the student to investigate this matter, and to determine for himself whether there is fair ground for the belief in directional influence.

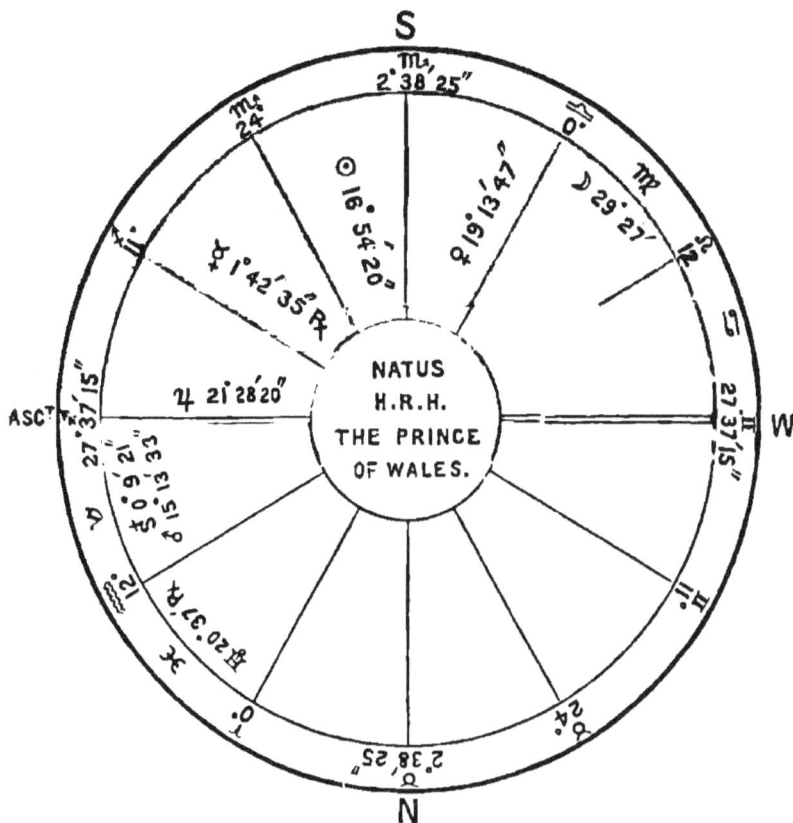

FIG. 7.

NATUS H.R.H. THE PRINCE OF WALES.

SPECULUM.

	Lat.	Dec.	R.A.	M.D.	S.-Arc.
	° ′ ″	16 54 5 s	224 26 6	13 59 53	67 32 32
☉	4 10 37 s	3 36 48 s	177 49 35	32 36 38	85 26 47
☿	1 46 29 s	22 15 38 s	239 12 44	28 46 31	59 1 43
♀	1 41 53 N	5 57 47 s	198 23 27	12 2 46	82 27 15
♂	1 32 36 s	24 9 37 s	286 43 19	103 42 54	124 19 51
♃	0 13 17 N	22 57 56 s	260 43 55	50 17 42	57 48 27
♄	0 43 44 N	22 43 56 s	270 10 18	120 15 55	121 47 6
♅	0 47 10 s	4 26 44 s	351 41 17	38 44 56	95 36 32

The horoscope of H.R.H. the Prince of Wales will prove an interesting one for such research. The official bulletin announcing his birth stated that it took place at 10ʰ 48ᵐ a.m. of the 9th November, 1841, at Bucking-ham Palace. The right-ascension of the meridian was then 14ʰ 1ᵐ 44ˢ.86 or 210° 26′ 13″. The latitude of the birthplace is 51° 30′ 3″ N., and the long. 0ᵐ 39ˢ W. of Greenwich. *Scorpio* 2° 38′ 25″ culminated, and *Sagittarius* 27° 37′ 15″ ascended, as will be seen in the preceding diagram.

Taking the time of birth given as correct, and it was evidently very carefully noted, the following directions correspond to the period of the Prince Consort's death; the arc for this early bereavement being 20° 6′ :—

		″	′
M.C. par. ♅ zod., con.	....	20	7
☽ 150° ♅ zod., eon.	....	20	10
Asc. ☌ ♂ m., d.	....	20	37

The marriage of the Prince of Wales took place on the 10th of March, 1863, and was celebrated with great pomp and national rejoicings, the public reception of his charming bride being a magnificent one. The arc for the marriage is 21° 20′, and the following directions measure very nearly the same :—

		°	′
♀ par. ☽ zod., con. (first contact)	...	21	22
♀ par. ☽ zod., con. (central)	....	21	47

In the month of November, 1871, H.R.H. became ill with typhoid fever, and in December his life was well-nigh despaired of. The are for this severe illness is 30° 1′. The direction of the Sun to the parallel decli-nation with Saturn was then operating; the Sun being " hyleg " and Saturn being in the ascendant of the nativity, this direction was a most dangerous one.

Other trying directions nearly coincided. Happily, the Sun was near enough to the par. dec. with Jupiter to save life (astrologically speaking):—

		°	′
☉ par. ♄ zod., d. (first contact) ...		29	43
♀ ♂ Ascendt. zod., d.		29	49
☉ par. ♅ zod., con.		30	6
Asc. ♂ ☿ m., con.		30	15

In October, 1876, the Prince went to India. The following directions were then operating:—

	°	′
☉ par. ♃ zod., d.	35	7
☉ △ ♄ m., con.	35	40

The ascending position of Jupiter at the moment of birth of H.R.H. the Prince of Wales, the sextile aspect between Jupiter and Venus, the square of the Moon with Jupiter, and the parallel declination of Mercury with Jupiter; the angular positions of the Sun (in the meridian) and Mars (in the ascendant) in mutual sextile; these striking and fortunate positions and configurations pre-signified the *bonhomie*, the good nature, and the great popularity of the heir to the British crown. The Prince is eminently fitted by nature for his high position and for that still higher state to which he will one day accede. Indeed, this horoscope tells strongly in favour of genethliacal astrology. The drawbacks are the angular position of Saturn (in the ascendant) and his square aspect with the Moon, which have already had effect in the dangerous illness suffered by the Prince in 1871, and in the widowhood of Her Majesty. The positions of the Sun and Mars are highly favourable for the victorious career of the army and navy of Great Britain, should war, under the Prince's sway, be, unhappily, unavoidable. In fine, the nation may anticipate for H.R.H. a most glorious and fortunate reign.

CHAPTER XIX.

ON SOLAR REVOLUTIONS.

"Vitam regit fortuna, non sapientia."—Cicero.

A Solar Revolution is the return of the Sun to the degree, minute, and second of longitude held by him at birth. A map of the heavens is drawn for the exact moment of the Sun's return. The Solar revolution is, of course, purely symbolical, and is of slight importance unless the same sign ascend as at birth, or unless a planet be exactly (or within 5°) rising, southing, setting, or in the lower meridian. When neither of these happens, it is advisable to reject the figure, and mark in pencil in the horoscope the transits at the revolution. Solar Revolutions derive most of their force from the transits, and these are then most potent because they happen at or near the period when the earth is in the same part of her orbit as at birth.

The moment of the Sun's return may readily be computed from either the *Nautical Almanac* or *Zadkiel's Ephemeris*. Ephemerides that do not give the Sun's longitude to *seconds* are useless for this purpose. The Sun's place at birth must be computed to seconds, or the moment of the Solar return cannot be exactly determined; in the case of a sign of short ascension being in the ascendant the revolution would be completely altered if there were an error of one minute in the Sun's longitude.

In the year 1870, the rulers of France and Prussia

had the Sun in conjunction with Mars at their Solar revolutions. In July of that year the Emperor of the French declared war against the King of Prussia. As the former suffered most by the result of that sanguinary war, so disastrous for France, it will be both interesting and instructive to examine his Solar Revolution for 1870, and to compare it with his horoscope (see page 100). By so doing we shall find a very singular and striking corroboration of the theory that the planets are "for signs of future events."

Here follows the calculation of the Sun's return:—

	°	′	″
Sun's long. at birth	29	44	55·0
Sun's long. April 19, 1870	29	12	20·7
	0	32	34·3

	°	′	″
Sun's long. April 20, 1870	30	10	51·2
Sun's long. April 19, 1870	29	12	20·7
Sun's motion in 24ʰ	0	58	30·5

Then, by proportional logarithms, say

As 58′ 30·5″	...............(a.c.)	9·51194
To 32′ 34·3	....	·74243
So 24ʰ 0	....	·87506
To 13ʰ 21ᵐ 40ˢ	................=	1·12943

The Solar Revolution takes place, then, at 1ʰ 21ᵐ 40ˢ a.m. (G.M.T.) of April 20th, 1870. This must be corrected for Paris mean time by adding 9ᵐ 21ˢ. Thus we find that the map of the heavens has to be drawn for 1ʰ 31ᵐ 1ˢ a.m., and for the latitude of Paris (48° 50′ N.).

The sidereal time at Greenwich mean noon of April

19th, 1870, is 1ʰ 49ᵐ 32·28ˢ.   From this must be subtracted 1·54ˢ, and we have

	h.	m.	s.
Sidereal Time at Paris	1	49	30·74
Add time elapsed	13	31	1
Add diff. mean and sidereal time	0	2	13·22

R.A. of M.C., 230° 41' 15" or 15 22 44·96

To find the ascending degree, we add 90° to the R.A. of the meridian, and thus obtain the oblique-ascension of the ascendant, viz. 320° 41' 15". Then, by Formula 1 (oblique-ascension given to find ecliptic longitude), p. 241 of vol. i. of the "Text-Book of Astrology," we determine the ascending degree to be ♑ 21° 53' 11". The same sign ascends as at birth.

The degree culminating is ♏ 23° 5' 11", which is almost the exact place of Saturn at birth, viz. ♏ 20° 24' ℞. The evil Uranus is just setting—being on the cusp of the house of war—in square aspect with Mars and in opposition to the ascendant. Mars is in exact square aspect with the ascendant! The Moon is in conjunction and the Sun in square with Saturn!

On the 8th of July, 1870, Uranus arrived at the exact opposition (♋ 21° 53') of the revolutional ascendant, and the dispute was then at its height between France and Prussia as to the Spanish succession. On the 14th of July, the Sun arrived at the same place (♋ 21° 53'), and the square of the place of Mars at the revolution; on the 12th, the Moon was totally eclipsed in the 21° of ♑, i.e. in the ascendant of the revolution, and this eclipse was visible in Europe; on the 15th, the Emperor of the French declared war against the King of Prussia. The decisive battle of Sedan began about 6ʰ a.m. of the 1st of September, 1870, when the Moon was in the place of Saturn at birth, and Mars was in conjunction with Uranus in the seventh house (that of war) of Louis Napoleon's horoscope and revolution. On the following day the Emperor surrendered to the King of Prussia, and became a prisoner of war. On the 4th of September the *déchéance* of the Empire was declared and the French Republic was proclaimed.

FIG. .

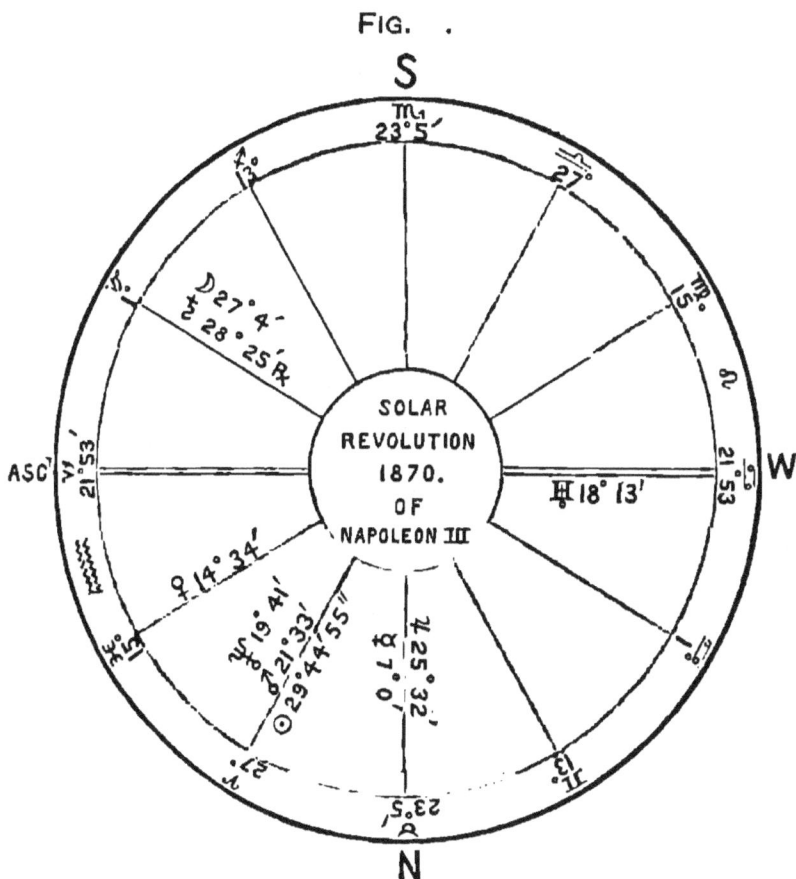

At birth the Sun was in conjunction with Mars; at the Solar Revolution, 1870, the Sun was again in conjunction with Mars, and the Emperor goes to war, meets with defeat, and loses his throne !

The King of Prussia had the Sun in conjunction with

Mars at his Solar revolution for 1870, but having a more fortunate horoscope (see *Urania* for April, 1880, p. 109), gains the victory over the unlucky Emperor of the French. These cannot be mere coincidences.

At H.M. Queen Victoria's Solar Revolution for 1854 (May 23rd, 3ʰ 28ᵐ 23ˢ p.m. G.M.T.), we find ♄ in 4° 48′ of *Gemini*, close to the places of the Sun and Moon and the ascendant at birth, and ♂ in ♍ 7° 43′, in *square* aspect to those places and ♄; the Moon in ♈ 25° 34′, close to the place of ♂ at birth. The Crimean War was begun that year.

At the late Czar's Solar Revolution for 1877 (April 28th, 4ʰ 21ᵐ p.m., Moscow), we find the Sun in square aspect with Mars. On the 24th of April, the Czar declared war against the Sultan of Turkey, on the very day that Mars passed through the degree of longitude (♒ 4° 42′) on the cusp of the house of war of his horoscope. (See *Urania* for February, 1880, p. 41.)

At the Solar Revolution of November 28th, 1882, 11ʰ 49ᵐ 42ˢ p.m., Madrid mean time, the King of Spain will have the Sun in conjunction with Mars, and the Moon in the place of Saturn at his birth (10ʰ 15ᵐ p.m., November 28th, 1857, Madrid). The Sun will be in ♐ 6° 42′ 33″, ♂ in ♐ 10° 5′, ☽ in ♋ 25° 49′. If he escape war in 1882-3, this young monarch will be in danger of a violent death. In September, 1882, he will have ☽ ☍ ♂ zod., con. 24° 52′, in operation; he will then be in danger in travelling.

As I am frequently asked for a copy of my horoscope, I will here state the time of my birth and give the planets' places thereat. My father (a legally qualified surgeon), registered my birth as occurring at 9ʰ 20ᵐ a.m. of November 10th, 1840 (in lat. 51° 30′ 35″ N., and long. 0ᵐ 31ˢ W.) This I have rectified to 9ʰ 18ᵐ 9ˢ a.m, which gives the R.A. of the M.C. 189° 8′ 10″;

♎ 9° 56' 37" being in the meridian, and ♐ 10° 13' 14" in the ascendant. ☉ in ♏ 18° 5' 31", ☽ ☍ 27° 6', ☿ ♐ 10° 29' 46", ♀ ♐ 16° 7' 38", ♂ ♍ 18° 4' 10", ♃ ♏ 26° 58' 46", ♄ ♐ 20° 5' 38", and ♅ ♓ 16° 34' 26" ℞. At my Solar Revolution for 1848, the Sun was in conjunction with Mars and Mercury, the Moon was in opposition to Mars, and Saturn was exactly in the lower meridian. My ninth year was, accordingly, a most critical one, my life was in the greatest danger, and my father suffered also.

Generally speaking, when the fortunes are found conjoined or favourably configurated with the Sun or Moon, and when Venus or Jupiter is in the ascendant or midheaven at a solar revolution, the succeeding year of life proves fortunate and healthy.

When Mars, Saturn, or Uranus is found in conjunction or evilly configurated with the Sun or Moon, and in one of the angles, the ensuing year of life proves either unfortunate or unhealthy, sometimes both.

Solar revolutions are not so potent when their features do not correspond with the nature of the primary directions operating at the period of their occurrence.

The Moon at a Solar Revolution in the place of Mercury at birth, and unafflicted, pre-signifies active and profitable employment, travelling, literary fame, etc., according to the strength of ☿ in the nativity. The Moon in the place of Venus, a pleasurable year. In the place of Mars, trouble and suffering, danger by fire, quarrels, etc., according to the nativity. In the place of Jupiter, gain, pleasure-seeking, and good health. In the place of Saturn, sickness, danger of accidents, pecuniary loss, etc. In the place of Uranus, a family loss, danger to wife, mother, or sister.

The effects of the transits of the planets are described in the "Text Book of Astrology," vol. i.

CHAPTER XX.

HORARY ASTROLOGY.

"They marked the influence, and observed the power
Of every *sign*, and every fatal *hour*."—MANILIUS.

HORARY ASTROLOGY consists of divination by means of the planets. The planets are used as symbols, the principle involved being that the human mind is constituted in deep sympathy with the movements and configurations of the heavenly bodies. Believers in horary astrology hold that the heavenly bodies, will, at any time of really anxious inquiry, pre-signify the connection and result of the circumstances affecting and surrounding the "querent"— thus forming a kind of oracle, and being, most probably the mode of divination employed by the ancients.

Niebuhr, in reference to the derivation of the Roman from the Etruscan religion, says: "Every department of divination was taught in the schools of the priests." Upon which Walter remarks: "This was, perhaps, not unconnected with astrology."

The *teraphim* mentioned in the Old Testament were images formed for the purposes of worship and for divining future events. They were made by astrologers under certain constellations. Among other reasons why Rachel stole her father's images, this is thought to be one, viz., that Laban should not, by consulting with them, discover which way Jacob took in his flight.

---

¹ See "Ecclesiastical Rites used by the Ancient Hebrews," by Thomas Goodwyn. 1628.

Joseph was acquainted with divination by horary astrology. His divining cup had engraved on its outer side the symbols of the signs of the zodiac, and the horoscope ready for the marking in of the positions of the planets at the moment of consultation.

This was the cup which was found in Benjamin's sack. Joseph sent his steward after his brethren with instructions to say to them: " Wherefore have ye rewarded evil for good ? Is not this [cup] it in which my lord drinketh, and whereby indeed he divineth ? " And when his brethren were brought back to him, Joseph said: " What is this that ye have done ? Wot ye not that a man such as I can certainly divine ? "

Allusions are frequently met with in the writings of modern Persian poets to the divining cups used by Persian monarchs; in fact, much of the wonderful prosperity of the ancient monarchs of Persia is attributed by the poets to " the cup showing the universe."

Diodorus avers that Joseph was the author of a great work on astrology: " The Aphorisms of Hermes the Egyptian."

The term "horary" is derived from the Latin word *hora*, an hour. The *hour* at which the querent consulted the astrologer was noted and a map of the heavens drawn.

Among the Singhalese, as among the early Arabian and European astrologers, Hora is a *planetary hour.* The twenty-four hours of the day were each in turn dedicated to the seven planets, in the following order, beginning with the first hour after sunrise on Sunday:—the Sun, Venus, Mercury, the Moon, Saturn, Jupiter, Mars. Thus the theory of the planetary hours gave origin to the nomenclature of the several days of the week, and to the division of the week into seven days—the number of the planets (the Sun and Moon being included under the head of planets). The seventh day was Saturn's day (Saturday), and this was *dies infaustus*, and accordingly all work was suspended on the seventh day of the week.

The order of the planetary hours was derived from the order of application. The Moon applies to Mercury (the Moon being swifter in motion than Mercury), Mercury to Venus, Venus to the Sun, the Sun to Mars, Mars to Jupiter, and Jupiter to Saturn. Hence, Saturn being the " highest " and slowest in movement of the planets known to the ancients, the order became: Saturn, Jupiter, Mars, the Sun, Venus, Mercury, the Moon.

Horus, or *Orus*, was, according to Herodotus, the Egyptian name for the Sun. The Sun, as already stated, ruled the first day and the first hour of that day of the week; hence the 25th hour, or the first of the next day, was the planetary hour of the Moon. The following table will explain this very simple arrangement, which Whewell, being unacquainted with astrology, mistook for " certain arbitrary arithmetical processes connected in some way with astrological views."

Whewell says: " The usage is found all over the East; it existed among the Arabians, Assyrians, and Egyptians. The same week is found in India, among the Brahmins; it has there also its days marked by the names of the heavenly bodies. The period has gone on without interruption or irregularity from the earliest recorded times to our own days, traversing the extent of ages and the revolutions of empires."

Eusebius gives an ancient oracle, which he had copied from Porphyry, in which we are clearly informed that the *Magi* named the days of the week from the planets.

TABLE OF THE PLANETARY HOURS.

Hours.	1 Sun. Sunday.	2 Moon. Monday.	3 Mars. Tuesday.	4 Mercury. Wednesday.	5 Jupiter. Thursday.	6 Venus. Friday.	7 Saturn. Saturday.	
1	☉	☽	♂	☿	♃	♀	♄	Sunrise.
2	♀	♄	☉	☽	♂	☿	♃	
3	☿	♃	♀	♄	☉	☽	♂	
4	☽	♂	☿	♃	♀	♄	☉	
5	♄	☉	☽	♂	☿	♃	♀	Hours of the Day.
6	♃	♀	♄	☉	☽	♂	☿	
7	♂	☿	♃	♀	♄	☉	☽	
8	☉	☽	♂	☿	♃	♀	♄	
9	♀	♄	☉	☽	♂	☿	♃	
10	☿	♃	♀	♄	☉	☽	♂	
11	☽	♂	☿	♃	♀	♄	☉	
12	♄	☉	☽	♂	☿	♃	♀	
1	♃	♀	♄	☉	☽	♂	☿	Sunset.
2	♂	☿	♃	♀	♄	☉	☽	
3	☉	☽	♂	☿	♃	♀	♄	
4	♀	♄	☉	☽	♂	☿	♃	
5	☿	♃	♀	♄	☉	☽	♂	Hours of the Night.
6	☽	♂	☿	♃	♀	♄	☉	
7	♄	☉	☽	♂	☿	♃	♀	
8	♃	♀	♄	☉	☽	♂	☿	
9	♂	☿	♃	♀	♄	☉	☽	
10	☉	☽	♂	☿	♃	♀	♄	
11	♀	♄	☉	☽	♂	☿	♃	
12	☿	♃	♀	♄	☉	☽	♂	

E

The early Christians changed the day of rest from the seventh (Saturn's day), as instituted by Moses, to the first day (Sun-day) of the week. This was the Druidical Sab-aith day. Many ages before Christianity —in the remotest times of Britain and Gaul—the first day of the week was set apart, more particularly, for the instruction of the laity, and distinguished by the name of " the day of the Sun." A Saronide, or preacher, delivered his sermon from a jube, or pulpit. The discourse was termed Sab-aith, or " the word of the teacher or wise man."

The Jewish sacrificial system as described in Numbers xxvii., and elsewhere, had reference to the motions and influences of the heavenly bodies. The morning and the evening sacrifice were obviously connected with the Sun; the Sabbath offering, with the planet Saturn; the offering of the new moon, with the pre-signification of future events attached to the new moon (or rather the planetary positions thereat); and the paschal sacrifice, with the vernal equinox and the lunation nearest to the Sun's passage of the equator at the rising sign of Aries.

The horary astrologer before giving judgment on the subject of inquiry examined the figure of the heavens, drawn for the moment of consultation, to determine whether it was " radical " (i.e. like the radix, or nativity of the querent); and if he found that the planet ruling the ascending sign was of the same triplicity as the " lord of the hour " of consultation, he deemed to be radical, or fit to be judged, whether or no the nativity was procurable. Thus, if a fiery sign (Υ, Ω, or $\nearrow$) ascended, and the Sun or Mars, in the daytime, was lord of the hour, the figure was deemed radical. To determine the planetary hour, during the day, the horary astrologer divided the time of sunset by 6, and by this means obtained the length of the planetary hour;

then from the hour of consultation he subtracted the time of sunrise, and thus readily ascertained (with the aid of the foregoing table) the planetary hour.

The planets Uranus and Neptune, not being known to the ancients, were not included in the table of planetary hours. The value of it, at the present day, is *nil*. The only guarantee that the horary figure is fit to be judged, is the deep anxiety of the querent and the resemblance of the figure to his nativity. It is only at moments of the profoundest anxiety—when life or fortune may be trembling in the balance—that horary figures are worth consideration. In most cases, the nativity, if procurable, would solve the questions with greater certainty.

The confident statements of Lilly and his *confrères* as to the reliability of horary astrology are really wonderful. They applied it alike to the most momentous and the most trivial affairs. The horary astrologer, without the slightest acquaintance with anatomy, physiology, and pathology, took upon himself to decide where doctors disagreed, and to diagnose the nature of an obscure disease which had baffled the wisest physicians.

In a work on horary astrology, published in 1861, an example was given of a figure taken for the moment of reading a newspaper report of the arrest of a young lady on the charge of murdering her little half-brother. The horary astrologer did not stop to reflect that a hundred astrologers residing in different parts of the United Kingdom would read the report at widely different times, and would, if their curiosity were excited, take horary figures at different hours, in not two of which figures might there be any real resemblance. The young lady was released, as he declared she would be; but her guilt was, many years afterwards, confessed by herself.

In the preface to Lilly's " Introduction to Astrology,"

we are told that horary astrology may be quickly learned by any person of even moderate abilities, and may be well understood and reduced to constant practice in three months. Also, that after many years' experience the author had found the laws of horary astrology unfailing. The great majority of his clients consulted him by letter, hence it was simply impossible for him to verify the correctness of his judgment in every instance. The same writer stated that if in a figure taken for the beginning of a serious illness the testimonies agree, " your judgment," as to the nature of the disease, " will be infallible." As the writer was not a medical man, and was never engaged in medical practice, he had not sufficient opportunities for testing the truth of horary astrology in cases of sickness. Certainly, to the eye of the trained physician the aphorisms of Lilly, in regard to illness, appear a hopeless jumble. The nativity of the patient would in most cases afford a good prognosis, but neither it nor the figure for the decumbiture could be of much use in diagnosis.

The rules of horary astrology which we are about to present to our readers, in a condensed form, are applicable only with safety to figures concerning events of the greatest importance.

CHAPTER XXI.

HORARY ASTROLOGY—Continued.

The Twelve Houses of the Heavens have the following signification in Horary Astrology.

The *Ascendant*, or First House.—This represents the person asking the question, the one who makes an offer to another person, the plaintiff in an action at law, etc. Should an evil planet be within 5° of the eastern horizon, the person will have a mark, blemish, or scar, either on the face or on that part of the body ruled by the sign occupied by the planet. For example, if ♂ be rising in ♈ the person would have a mark or scar on the face, near the eyes; in ♉, on the neck.

The *Second House* has relation to property in goods, money, and chattels; in law-suits it denotes the plaintiff's friends or assistants, witnesses, etc.; in battles, the support or aid a general may expect.

The *Third House* governs brethren, sisters, near kindred; short journies, and removals; letters, rumours, messages, newspapers; and railways.

The *Fourth House* has relation to the father of the querent; to houses, lands, mines, inheritance, and hidden treasures.

The *Fifth House* bears relation to children, schools, theatres, and places of amusement.

The *Sixth House* concerns servants, small cattle, and sickness.

The *Seventh House* relates to marriage, love engagements, the wife or husband; and (in cases of crime) criminals.

The *Eighth House* has relation to deaths, legacies, wills; to the wife's dowry, or the husband's estate.

The *Ninth House* relates to voyages, long journies, religion, science, art, literature, books, beneficed clergy, etc.

The *Tenth House* rules the monarch, president of a republic, or prime minister; business and commerce generally, and professions.

The *Eleventh House* represents the querent's friends, his hopes and desires. To kings and governments it concerns their friends, allies, &c.

The *Twelfth House* relates to the querent's secret enemies; to great cattle; to prisons and prisoners.

Bearing in mind these distinctions, it is easy to determine to which house belongs the affair, the issue of which it is desired to ascertain.

But there is another thing to consider before proceeding to judgment, viz., to which *planet* the affair inquired about relates. The following statement of the persons, professions, commodities, etc., ruled by the several planets, will serve to this end :—

The SUN.—*Professions*, etc.; Monarchs, princes, prime ministers, dukes, and high dignitaries generally, whether in nations, cities, towns, or armies; also goldsmiths, braziers, coppersmiths, minters of money and pewterers. *Mineral :* Gold. *Stones :* Ruby, carbuncle, hyacinth, and chrysolite.

The MOON.—*Professions*, etc.: Ladies of title and quality, whether queens, princesses, or countesses; women generally. Also watermen, travellers, pilgrims, fishermen, fishmongers, vintners, brewers, sailors, and hawkers. *Mineral :* Silver. *Stones :* The crystal, the selenite, and all soft white stones.

MERCURY.—*Professions,* etc.: Astronomers, astrologers, philosophers, poets, editors, journalists, accountants, attorneys, clerks, schoolmasters, tailors, most mechanics, and money-lenders. *Mineral:* Quicksilver (named mercury). *Stones:* Topaz, and all stones of various colours.

VENUS.—*Professions*, etc.: Musicians, actors and actresses, artists, engravers, painters, gamesters, linen-drapers, perfumers, and all who deal in ornaments, women's apparel, and toys. *Minerals:* Copper and white metals, sometimes silver. *Stones:* White coral, chrysolite, light sky-coloured sapphire.

MARS.—*Professions*, etc.: Surgeons, apothecaries, chemists, soldiers, gunners, barbers, cutlers, cooks, smiths, ironmongers, and all who work with sharp tools. *Minerals:* Iron, antimony, arsenic, brimstone, and ochre. *Stones:* Bloodstone, loadstone, jasper, firestone, and all common red stones.

JUPITER.—*Professions*, etc.: Bishops, the clergy, judges, barristers, students at universities or colleges, scholars, clothiers, woollen drapers, grocers and their assistants. *Mineral:* Tin. *Stones:* Amethysts, crystals, emeralds, hyacinths, marble, sapphire, topaz and tur-quoise; also all stones of blue colour or red mixed with green.

SATURN.—*Professions*, etc.: Popes, monks, friars, hermits, ascetics; agriculturists, brickmakers, chandlers, dyers, miners, sextons, gravediggers, and shepherds. *Minerals:* Lead, coal, and all dross or earth. *Stones:* All black or ash-coloured stones.

URANUS.—*Professions*, etc.: Astronomers, astrologers, philosophers; and all who experiment in chemistry, electricity, &c.

Having assigned the affair inquired about to its proper planet and house, it next becomes necessary to determine the planet ruling the inquirer (his " signifi-cator "). The planet whose sign is in the ascendant is usually considered the querent's significator, and the Moon as his or her co-significator (except in a love affair, when the Moon is always taken as the co-significator of the man, and the Sun as the co-significator of the woman).

The following table of the "houses" or chief dignities of the planets is exactly identical with that found in the mummy case of the Archon of Thebes, in ancient Egypt.

♌	☉	☽	♋
♍	☿		♊
♎	♀		♉
♏	♂		♈
♐	♃		♓
♑	♄		♒

The Sun has but one "house," viz. *Leo*. The Moon has only one, *Cancer*. The planets Mercury, Venus, Mars, Jupiter and Saturn have two each. The signs opposite to their chief dignities are those in which the planets suffer "detriment," as, for instance, the Sun in *Aquarius*, the Moon in *Capricornus*, Mercury in ♐ and ♓. In certain signs the heavenly bodies are said to be exalted. The Sun's exaltation is in ♈, the Moon's in ♉, Mercury's in ♍, Venus's in ♓, Mars's in ♑, Jupiter's in ♋, and Saturn's in ♎.

Uranus and Neptune not being known to the ancients were left out in the cold. As they are more remote from the Sun than is Saturn, they should have the same houses and exaltation as ♄.

The industrious character of most people born with ♍ rising, the fiery nature of most persons born under ♈ and ♏, and the joviality of ♐ and ♓ persons, would appear to justify the above arrangement of chief dignities.

Should Pisces ascend at the moment of consultation,

Jupiter would be taken as "lord of the ascendant," or ruling planet. At one time this rule was applied to nativities, but it is now very properly discarded by experienced students, and confined to horary and mundane astrology.

Let us now pass on to the consideration of affairs ruled by the *first* house of the Heavens.

1. Whether an event suddenly occurring will prove fortunate or unfortunate?

Note the moment of occurrence of an important event which excites anxiety or fear, and draw a map of the Heavens for such moment. If the ascendant have ♃ or ♀ located therein, unafflicted, if ☽ be well aspected and not afflicted, and no evil planet be in the midheaven or evilly configurated with degree ascending, then no trouble will accrue from the event. On the other hand, if an evil planet be in the ascendant, or midheaven, the ☽ afflicted, and the planet corresponding to the nature of the event, also afflicted, then no good will come of the affair, and it had better be abandoned or let alone if possible. An evil planet in the second house afflicting the ☽ or planet in the ascendant, pre-signifies pecuniary loss; in the third, trouble through a relative or neighbour, or short journey; in the fourth, trouble through the father, or concerning land, or houses, or mines; and, in like manner, judge respecting the other houses. A benefic so situated and in good aspect with the ☽ would show gain.

2. Of the Fate of a Ship.

The figure of the Heavens must be drawn for the moment of the launch of the vessel, if it can be procured. Figures drawn for the commencement of a voyage are unreliable; for many vessels set sail at or very near the same moment, yet some arrive safely at their destination and make remunerative voyages, while others meet with misfortunes and losses.

The ascendant and the ☽ are taken to signify the ship.

An evil planet in the ascendant, or afflicting the ☽ or the ascendant, the ship will be unfortunate. If the afflicting planet be ♄, the vessel will run aground. If ♂, very great danger or damage (by fire, probably, especially in a fiery sign). If the signs containing the afflicting planets be such as rule the parts of the ship below water, a leak will be sprung. If ♂ and ♅ be conjoined in the fourth house, explosion is to be feared. If ♃ or ♀ be in the ascendant, the second house, or the meridian, the ship will prove a profitable investment. An evil planet in the second house, there is every probability that she will prove a losing venture. The parts of the ship ruled by the various signs are: ♈, the breast or bows; ♉, the cutwater and parts beneath; ♊, the rudder and stern; ♋, the bottom; ♌, the upper works; ♍, the hold; ♎, the parts about the water's edge; ♏, the seamen's berths, and cabin; ♐, the crew; ♑, the ends of the vessel; ♒, the master or captain; ♓, the wheels or screw in steam-vessels, and the sails in sailing ships.

While on this subject it may be interesting to recount the following incident. On the 6th of November, 1877, the author was in company with a gentleman who was part owner of a steamship which was launched that day at 45 minutes past noon at Port Glasgow. I drew a map of the Heavens for the moment of launching. The R.A. of the meridian was 15^h 48^m 20^s; the 29th deg. of ♏ was in the midheaven, the Moon being in ♏ 29° 20' and in square aspect with Uranus in ♌ 29° 7', in the seventh house. The 19th deg. of ♑ was in the ascendant and the infortunes Mars and Saturn, nearly conjoined, were in the first house (in ♓). I told Mr.

C. (the part owner) that the vessel—the Kerangie— would soon meet with a disaster, and that she would prove anything but a profitable investment. On her trial trip she met with an accident; and not many months afterwards she struck on a rock while on a voyage in the southern ocean. She proved a most unprofitable investment.

It would be well for owners if they would select a fortunate moment for the launching of their ships.

CHAPTER XXII.

HORARY ASTROLOGY—Continued.

Next in order come the questions pertaining to the *second* house or division of the Heavens.

1. Whether the querent shall have money lent returned to him?

The lord of the ascendant and the ☽ are for the querent, the lord of the second (or planet in the second) denotes his substance. The lord of the seventh is the significator of the quesited, and the lord of the eighth (or planet in the eighth) of his substance. But, if the quesited be a relation, then the house ruling such relationship must be taken for him—as, for example, the third for a brother or sister, the fourth for the father.

Testimonies of the return of the money: Lord of ascendant or ☽ in conjunction or good aspect with the lord of the eighth or planet in the eighth. The lord of the eighth house in the second, with reception. Lord of ascendant or ☽ joined to a fortune having dignities in the ascending sign; or in the tenth house.

Testimonies of the loss of the whole or part of the money: If the planet in the eighth house be an infortune, and have no reception with either of the querent's significators, the money will only be recovered in part. Lord of the seventh or eighth in the first or second house, without reception with either of the significators of the querent or his substance, the money will never be recovered.

Note.—If the time of recovery be part of the question, count the number of degrees between the platic and perfect aspect between the significators, and say that every degree in moveable signs and angles answers to a day; in common signs and angles, a week; in fixed signs and angles, a month. Succedent houses give weeks, months, and years, according to the sign; cadent houses show months and years.

2. Whether the querent shall obtain preferment or employment from the Government, or person of high rank, or an employer?

This must be judged in a similar way to the last question, except that the tenth house must be taken for the quesited. In all cases, see that the fourth house and its lord be not greatly afflicted, as the affair would then *end* badly.

Questions relating to the *third* house.

1. Of news, rumours, etc., whether importing good or evil to the querent?

Only when the report concerns the querent, his family, or his substance, should these rules be applied:—

Good news is true if ☽ be in the ascendant, third, tenth, or eleventh house, separating by good aspect from any planet, and applying by good aspect to the lord of the ascendant at the time of receiving the news.

The news is false if ☽ be void of course (*i.e.*, passing out of a sign without forming an aspect with any planet), or if ☽ be in evil aspect with ☿, and neither in good aspect to the lord of the ascendant.

2. Whether it will be to take a certain journey?

Only journeys that can be accomplished within 24 hours are ruled by the third house.

If the lord of the ascendant be strong and in the third house, or in reception with the lord of the third, or well aspected by it, or if the ☽ be so placed, the journey may be safely undertaken. A square aspect (in signs of short ascension), with reception, is not against the journey. If, on the other hand, there be no connection between the lord of the ascendant, or the ☽, with the lord of the third, and the lord of the ascendant, or ☽ be afflicted by an evil planet in the third, then it will *not* be well to take the journey.

Questions concerning the *fourth* house.

1. Of buying or selling houses, land, farms, etc.

The ascendant and its lord are for the purchaser, the seventh and its lord for the seller. The fourth house and its lord are for the house or land, etc. The tenth house and planet or planets therein are for the price of the house or land, and show whether the price is cheap or dear.

The lord of the ascendant and seventh in good aspect, the lord of the seventh applying to the lord of the ascendant, a bargain may be readily concluded. If the significators be in opposition without reception, there will be no bargain.

The quality of the house, or land, etc., may be known by the planet in the fourth, the sign therein and its lord. If two infortunes be therein, and the lord thereof be retrograde or afflicted, the property will not continue long in the purchaser's family. But if either ♀ or ♃ be in the fourth, it will be a profitable investment and will remain in possession of the family.

If there be an infortune in the ascendant, the occupiers are deceitful, and will not do justice to the house, or farm. If the infortune be direct in motion, they will purloin the timber or wear out the land, injure the buildings, etc. If it be retrograde, the occupiers will run away, or throw up their lease.

If a fortune be in the ascendant, the tenants will act honestly and give every satisfaction.

If the lord of the tenth house be strong and angularly

posited, the price is high; if it be weak and afflicted, the price is not too high.

2. If it be well to take a certain farm, house, or land, etc. ?

The ascendant and its lord are for the person who would hire; the seventh and its lord for the owner; the tenth is for the profit by the transaction; and the fourth house is for the end of the affair.

If the lord of the ascendant be in the first house, or if a fortune be therein, the farmer desires greatly to take the farm, and will find it a good bargain; but if an infortune be in the ascendant, he will go back from his bargain.

If the lord of the seventh be therein, or a fortune be therein, the man who has to sell, or let, will keep his word and abide by the bargain; but he will profit most by it.

A fortune in the fourth house, and well aspected, the affair will end well and satisfy all parties concerned. If an infortune be therein, the end of the transaction will not please either party.

3. If it be well to remove from one house or place to another ?

The ascendant and fourth house are for the place or house occupied by the querent; the seventh the place or house to which he would remove.

The ascendant or second house afflicted show that the querent is unfortunate or poor in his present abode; and, if the affliction be great and from fixed signs ($\mathcal{O}$, $\mathcal{Q}$, M, and $\approx$), he will never do any good there. If the fourth house be afflicted by evil planets, the house he occupies is unlucky or unhealthy, and he had better leave it.

The seventh and eighth houses show the result of his removal. If they be stronger or better aspected than the first, second, and fourth, he will do better in the place or house to which he thinks of removing. If they be weaker and more afflicted than the first, second, and fourth houses, then the querent would not gain by removing to such place or house.

Questions pertaining to the *sixth* house.

Whether a sickness will be long or short?

The time of the first attack of the illness should be obtained, if possible; as this will show the probable issue, and the critical periods.

Generally speaking, if a common sign (Π, $\mathfrak{m}$, $\cancel{\jmath}$, or $\mathcal{H}$) be in the sixth house, the sickness will be of its average duration; if a fixed sign ($\mho$, Ω, $\mathfrak{m}$ or $\approx$), the disease will be long and hard to cure; if a moveable sign (Υ, $\mathfrak{S}$, $\triangle$, or $\mathfrak{V}\mathfrak{f}$) be in the sixth, the sickness will be short; and, if the last few degrees of a sign be on the cusp of the sixth, a change for the better or worse is at hand, and the issue will quickly determine.

A benefic planet in the sixth (and not ruler of the sixth), dignified, the sickness will soon be over and recovery complete.

The lord of the ascendant and $\mathbb{D}$ well placed, swift in motion, and well aspected, the disease will be brief and recovery ensue.

Both luminaries cadent, their dispositors unfortunate, and the lord of the ascendant afflicted by $\mathfrak{h}$, or $\mathfrak{h}$ in the sixth in a fixed sign, expect a long illness, and one difficult to cure.

The lord of the ascendant or $\mathbb{D}$ in the sixth house, afflicted by the rulers of the fourth, sixth, and eighth houses, the sickness will be long, and the cure delayed.

All the significators of the sickness in fixed signs, the patient will be long ill; and if $\mathbb{D}$ be afflicted and applying to the $\mathcal{d}$ or evil aspect of the lord of the eighth, the illness is of a severe if not dangerous type.

The *crises* may be known by watching for the times when the ☽ reaches the semi-quartile, the quartile, opposition, and conjunction with her own place in the figure for the decumbiture, according to the nature of the illness. If when the ☽ reaches one of these points she be at the same time in sextile or trine aspect with Jupiter, Venus, or the Sun, there will be a change for the better; but, if ☽ be in ☌ or evil aspect with an infortune, the crisis will be an unfavourable one.

EXAMPLE.—Dr. H. was seized with expectoration of blood at 6ʰ a.m. of May 16th, 1865 (London). R.A. of M.C., 21ʰ 35ᵐ 51ˢ. Ascendant ♊ 27°. Uranus rising in ♊ 27° 39'. The Moon in ♑ 29° 31', in opposition to Mars in ♋ 25° 16', and in square with Saturn in ♎ 24° 56'ʀ. He died of pulmonary phthisis on the 25th of September, 1865, when Saturn arrived at the exact square aspect of the Moon's place at the first attack of hœmoptysis. The Moon was afflicted by the lord of the sixth house, Mars, and the lord of the eighth, Saturn. *Scorpio*, a fixed sign, was in the sixth house, and its lord (♂) had dignities in the eighth (♑). The Moon had no adequate assistance from the fortunes. The lord of the ascendant (☿) was retrograde in a fixed sign (♉). Mars, the ruler of the sixth house, was in *Cancer*, the sign that rules the chest.

CHAPTER XXIII.

HORARY ASTROLOGY—Continued.

We now come to the questions relating to the *seventh* house, and to marriage.

Whether the querent shall marry a certain person ?

The ascendant and its lord are for the querent; the descendant and its lord are for the quesited A planet within 5° of the horizon becomes a co-significator. The ☉ is for the lady, the ☽ is for the man.

If the lord of the ascendant dispose of the lord of the seventh, the querent is beloved by the quesited, and *vice versâ.* If there be mutual reception between the lords of the ascendant and descendant, the love is mutual.

If the lord of the first house or the ☽ apply to ☌ or good aspect with the lord of the seventh, or planet therein, the match may be effected by the querent's exertions. If there be translation of light between the significators, the match may be brought about by the intervention of a friend of the nature of the planet which translates the light. If the significators be in angles, and in good aspect by application and in mutual reception, the Sun and Moon also applying to mutual benefic aspect, and no evil planet interpose by quartile or opposition ray, the match will certainly take place.

An additional testimony is the good aspect of Venus to either of the significators, if she be in an angle and free from affliction. If Venus be much afflicted, no good will come of the affair.

If there be no good aspects between the significators, and if they be afflicted, in no reception, and unaided by Venus, the match will not be effected.

If the significators be all *separating* from good aspects, and Venus be afflicted, there will be no marriage. If the testimonies be contradictory there will be no immediate marriage, and the question may be deferred to a future day if the querent should still desire to bring it about.

If there be signs of marriage, yet some evil planet impede it, see what house such planet rules; if it be the second, pecuniary difficulties will interpose; if the third, the querent's kindred object; if it be the fourth, either the father of the querent or the mother of the quesited will interfere to prevent it.

Questions relating to the *ninth* house.

Whether a certain voyage will be safe, and advantageous to the querent?

If the lord of the ascendant be strong and in good aspect with the lord of the ninth (or with a benefic in the ninth), and if the lord of the ninth be well placed, the contemplated voyage will be both safe and advantageous. But, if Saturn or Mars be in the ninth house, and especially if at the same time in evil aspect with the lord of the ascendant or the ☽, the voyage had better be deferred. The ☽ in the ninth house and afflicted is a testimony of danger or loss. A benefic in the tenth or fourth house is a good testimony. If the fourth be afflicted, the voyage will end badly.

Questions relating to the *tenth* house.

1. Whether the querent shall obtain employment or office?

If the lords of the ascendant and of the tenth, and the Sun and Moon be well aspected and in mutual reception and strong, the querent will obtain the employment or office he is seeking. The lord of the tenth house located in the ascendant and in good aspect with the lord of the ascendant or ☽, presignifies success.

The lord of the ascendant in the tenth and well aspected, is a favourable testimony. The ☽ separating from the lord of the tenth and applying to the lord of the ascendant, is very favourable. Any assistance from the lord of the fourth, or from a benefic in an angle, is a token of success. Malefics, if strong and in good aspect with the lord of the ascendant, show success, but with some difficulty or delay. Benefics promise speedy success. If an evil planet afflict the lord of the ascendant, or planet in the ascendant, or the ☽, without reception, the employment will not be obtained.

2. Whether the querent shall continue in his employment or office?

This question may be judged in the same way as the last. The ☽ in the tenth and well aspected, or in good configuration with the lord of the tenth, especially with reception, the querent will continue in his employment.

Fixed signs in the ascendant and tenth are favourable; moveable signs are unfavourable. The lord of the ascendant cadent, in a moveable sign, and afflicted, the querent will hardly retain his employment.

EXAMPLE.—On the 3rd of January, 1878, at 4ʰ p.m. (London), the author received a letter from a friend who had been long out of employment and who had nearly exhausted his means in seeking some. The figure was *radical* for the same sign and almost the same degree ascended as at the querent's birth, and his great anxiety was unquestionable.

Upon casting the figure, R.A. of M.C., 22ʰ 52ᵐ 34ˢ, the student will find ♋ 14° in the ascendant; ☽ in the western angle, in ♑ 14° 10′, and hastening to conjunction with ♃ in ♑ 14° 44′; ♄ in the tenth, in ♓ 15° 34′. Finding that the ☽ wanted but half a degree to complete her conjunction with ♃, in a moveable sign and angle, the author gave it as his opinion that if there were any truth in horary astrology, the querent would obtain employment within a fortnight. This forecast was exactly verified, for on the 12th of January the querent obtained an appointment, to his great surprise and joy.

This sketch of horary astrology must now be brought to a conclusion. The author desires it to be distinctly understood that he deprecates the application of horary astrology to idle questions ; to impress on his readers that it is only applicable to affairs of the greatest moment and at times of the deepest anxiety, and to the moment of occurrence of events.

Bearing in mind the low ebb to which astrology sank, owing to the absurd and unwarrantable use made of the horary branch of it, in the middle ages, as satirised in "Sidrophel," it is impossible for the lover of truth for its own sake to wish that it should ever again become "popular" in that sense. In the hands of the intelligent and highly educated portion of the community, and of philosophers, astrology in all its branches might be made of the highest utility, and could be greatly improved.

It is a mistaken notion that the votaries of astrology, at the present day, are to be found almost entirely among the uneducated classes. The one hundred and fifty thousand purchasers of *Zadkiel's Almanac* belong chiefly to the educated portion of the public. A moment's consideration given to the contents of that popular *Almanac* will serve to show that they are utterly beyond the comprehension of the illiterate. The best writers on astrology do not pander to the tastes of the vulgar, nor encourage the superstitions of the ignorant. The fact is that knowledge of astrology and an intelligent faith in it spread concurrently with education.

Those of my readers who wish to pursue the science will find every aid afforded to them in " The Text-Book of Astrology."

APPENDIX.

ASTROLOGICAL VOCABULARY.

ACCIDENTAL POSITIONS.—These are the positions of the Sun, Moon, planets, and fixed stars in the various "houses" of the heavens at a birth, solar ingress, new moon, eclipse, etc. They are *mundane* positions. A planet in the ascendant, descendant, or either meridian, is said to be dignified "accidentally."

Affliction.—When the Sun, Moon, or a planet is in conjunction, parallel declination, semi-quartile, quartile, sesquiquadrate, or opposition with an evil planet, it is said to be afflicted. An evil planet within 5° of the ascendant or meridian, or in any evil aspect to either angle, is said to afflict it.

Airy Signs.—Gemini, Libra, and Aquarius, form the *airy* triplicity.

Altitude.—The angular distance of a heavenly body from the horizon, measured in the direction of a great circle passing through the zenith.

Anareta.—The planet that destroys life.

Angles.—The cardinal points are the commencement of the angles, which are the ascendant, upper meridian, descendant, and lower meridian.

Annual Variation of the right ascension or declination of a star is the change produced in either element by the effect of the precession of the equinoxes and proper motion of the star taken together.

Anomalistic Period.—The time of revolution of a primary or secondary planet in reference to its line of apsides. In the case of the earth, this period is termed the anomalistic year; in that of the Moon, the anomalistic month.

Aphelion.—That point in the orbit of a planet or comet which is most distant from the Sun, and at which the angular motion is slowest.

Apogee.—That point in the Moon's orbit which is furthest from the earth; the point in the earth's orbit which is furthest from the Sun. The greatest distance of any heavenly body from the earth.

Apparent Motion.—The motion of the celestial bodies as viewed from the earth.

Apparent Place of a star for any day is the position it appears to occupy in the heavens, as affected with abberation and nutation.

Apparent Noon.—The moment that the Sun's centre is on the meridian of a place.

Apparent Obliquity.—The obliquity of the ecliptic affected with nutation.

Application.—The motion of any celestial body towards the conjunction or aspect of another.

Apsides, Line of.—The imaginary line joining the aphelion and perihelion points in the orbit of a planet.

Aquarius.—The eleventh *sign* of the zodiac, which the Sun enters on or about the 21st of January. It is one of the ancient zodiacal constellations.

Arc.-—The distance between any two points in the heavens.

Arc of Direction.—The arc described by a planet when directed to another. The measure thereof is usually termed the *arc of direction.* Also the measure of the distance between any two points in the heavens, expressed in degrees and minutes.

Arc, Diurnal.—That part of a circle parallel to the equator described by a celestial body from its rising to its setting.

Arc, Nocturnal.—That part of a circle parallel to the equator described by a celestial body from its setting to rising.

Argument is a term used to denote any number or quantity by means of which another may be found.

Aries.—The first *sign* of the zodiac, which the Sun enters at the vernal equinox on the 21st of March. It is one of the ancient zodiacal constellations. The commencement of this sign, termed the *first point of Aries*, is the origin from which the *right-ascensions* of the heavenly bodies are reckoned on the *equator*, and their *longitudes* on the *ecliptic*.

Ascendant.—The eastern angle or first "house."

Ascension, Oblique.—The oblique ascension is the arc of the equator between the first point of *Aries* and the point of the equator which rises with a celestial body, reckoned according to the order of the signs.

Ascension, Right.—The distance of a heavenly body from the first point of *Aries*, measured upon the equator.

Ascensional Difference.—The difference between the right and oblique ascension.

Asteroids.—A name proposed by Sir W. Herschel for the minor planets between the orbits of Mars and Jupiter. They are chiefly out of the zodiac, and for this reason are not considered to have any influence in nativities. Still, when they happen to be in conjunction with either the Sun or Moon near the ecliptic, it might be advisable for the student to watch for any effects.

Aspects.—Certain distances between any points in the heavens. Those generally found to have force are : 18°, 24°, 30°, 36°, 45°, 60°, 72°, 90°, 108°, 120°,

135°, 144°, 150°, and 180°. The *conjunction* (when two heavenly bodies have the same longitude), and the *parallel declination* are usually included in the term "aspects." See Kepler's definition of aspects, p. 4.

Aspects, Benefic, are the following: the vigintile (18°), quindecile (24°), semi-sextile (30°), decile (36°), sextile (60°), quintile (72°), tredecile (108°), trine (120°), biquintile (144°). The quincunx (150°) was supposed by Kepler to be a good aspect, but experience teaches that it is only good when formed by a good planet, and that when formed by an evil planet, it is evil, like the ♂ and par. dec. The sextile and trine are the most powerfully good aspects.

Aspects, Malefic, are the following: the semi-quartile (45°), the quartile (90°), the sesquiquadrate (135°), and the opposition (180°).

BARREN SIGNS.—The signs Gemini, Leo, and Virgo.

Benefics.—The planets Jupiter and Venus ; also the Sun when strong and well aspected.

Besieged.—A planet found between two others. If between ♃ and ♀, it is fortunate; if between ♄ and ♂, ♂ and ♅, or ♄ and ♅, it is very unfortunate.

Bicorporal Signs are Gemini, Sagittarius (the first half of), and Pisces.

Biquintile.—The aspect of 144°.

CADENT.—Falling from angles. Planets in the third, sixth, ninth, and twelfth houses are cadent.

Cancer.—The fourth *sign* of the zodiac, which the Sun enters about the 21st of June, and one of the ancient zodiacal constellations. The beginning of the *sign* Cancer, 90° distant from the first point of *Aries*, is called the *Summer Solstice.*

Capricornus.—The tenth *sign* of the zodiac, which the Sun enters about the 21st of December, and one of the ancient zodiacal constellations. The beginning of the *sign* Caprieornus, 270° from the first point of *Aries*, is called the *Winter Solstice*.

Cardinal Points of the Ecliptic.—The equinoctial and solstitial points, *viz.* the first point of *Aries* and *Libra*, and of *Cancer* and *Capricornus*.

Cardinal Signs.—Aries, Cancer, Libra, and Capricornus.

Climacterical Years.—The years in life corresponding to the place of the Moon on those days after birth when she arrives at the *square* (90°) or *trine* (120°) of her place at birth.   They are the seventh, ninth, fourteenth, eighteenth, twenty-first, twenty-seventh, twenty-eighth, thirty-fifth, thirty-sixth, forty-second, forty-fifth, forty-ninth, fifty-fourth, fifty-sixth, sixty-third, and seventieth.   The forty-ninth and sixty-third are held to be the most important.   Great changes are frequently observed in these years.

Colours of the Planets, etc.—In *horary* astrology, the Sun is yellow; the Moon white, or silver; Mercury, light blue or striped; Venus, white; Mars, fiery red; Jupiter, red mixed with green; Saturn, black. Uranus and Neptune have no colours assigned to them, at present.   The signs of the zodiac have colours assigned to them, thus: ♈ white and red; ♉ red mixed with citron; ♊ red and white mixed; ♋ green or russet; ♌ red or green; ♍ black spotted with blue; ♎ black or dark brown; ♏ dark brown; ♐ light green or olive; ♑ black or very dark brown; ♒ sky blue; ♓ pure white, or glistening.   These colours were made use of to describe the dress of the quesited in horary questions.

Combust or *Combustion.*—A planet was said to be combust, or burnt up, when very near the Sun, especially within 5°.   In nativities it was held that the Sun takes the character of the planet, which loses its power. In horary questions, when the significator of the querent, or quesited, is within 5° of the Sun, it was thought to show a sickly and unfortunate person; and in all things it was considered a very evil testimony. Later experience shows that a planet within 5° of the Sun does not lose its power—in nativities.

Common Signs.—Gemini, Virgo, Sagittarius, and Pisces.

Configurations.—The relative positions of celestial bodies.

Conjunction.—Two heavenly bodies are said to be in conjunction when they have the same longitude. In astronomy they are also said to be in ☌ when they have the same right-ascension. The Moon is in conjunction with the Sun at the time of new Moon, both luminaries having then the same longitude—this is the *ecliptic* conjunction. In judicial astrology the conjunction is considered good with benefic planets, and evil with Mars, Saturn, and Uranus. But the ☌ of ♀ and ♂ is not evil, because these planets are "friendly."

Constellation.—A number of stars included within an imaginary figure in order that they may be the more easily identified. There are forty-eight constellations, which were formed more than two thousand years since, and are all referred to by Claudius Ptolemy in his great work called the "Almagest." These are usually called the ancient constellations. Others were introduced by Hevelius, and more recently M. de Lacaille has added a considerable number to fill up vacant spaces in the southern heavens.[1]

---

[1] See the description of the constellations given in the "Introduction to Astronomy" by Mr. J. R. Hind, F.R.A.S. Also the chapter on "Signs and Constellations of the zodiac," in the first volume of the "Text-Book of Astrology."

Converse Motion.—When the significator appears to move from east to west by reason of the rotation of the earth.

Co-significators.—The ☽ is the co-significator of the querent, in *horary* astrology, in all questions except those relating to love and marriage. See p. 146.

Copernican System.—The received theory of the Universe. More correctly the *Pythagorean* system. Its main facts were taught by Pythagoras, a celebrated Greek philosopher, who flourished about 500 years before the Christian era. Mr. Hind says: "It is even probable that some of the principles of the received system were current among the ancient Egyptians, a nation which distinguished itself for its acquaintance with the science of astronomy." In this system the Sun occupies the central place, and the planets with their attendant satellites revolve about him. Because this system was revived by Copernicus, a Prussian astronomer, in the sixteenth century, it is called the *Copernican* system ; nevertheless, the honour properly belongs to Pythagoras. The fact that Pythagoras believed in astrology is *concealed* by modern astronomers who are ever ready to repeat the parrot-cry that it was the Copernican system that overthrew astrology. See p. 22.

Critical Days.—Those on which the Moon forms a major aspect with her own place at the moment of a patient being first seized with illness. See p. 145.

Culminate.—To reach the meridian.

Cusp.—The commencement of any one of the twelve houses of the heavens.

Cycle.—A term generally applied to an interval of time in which the same phenomena recur.

Cycle of Eclipses.—A period of about 6,586 days, which is the time of a revolution of the Moon's node: after the lapse of this period, eclipses recur in the same order as before, with few exceptions. This cycle was known to the ancients under the name of *Saros*.

Cycle, Solar.—A period of 28 years, after which the days of the week correspond in the same order to the days of the month.

Decanate.—The signs were divided into three equal parts of 10° each, beginning with ♈ 0°. The first 10° of ♈ belonged to ♂, the second decanate to the ☉, the third to ♀. Each following decanate was assigned to a planet in the order following: viz. ☿, ☽, ♄, ♃, ♂, ☉, ♀. This is the same order as that of the planetary hours (see p. 129). A planet in its own decanate was considered to be dignified.

Decile.—The aspect of 36°.

Declination.—The angular distance of a heavenly body from the equator, either north or south. The complement of the declination is termed the *polar distance.*

Decumbiture.—A map of the heavens drawn for the moment at which a sufferer first goes to bed ill, is termed the "decumbiture."

Degree.—In the sexagesimal division of the circle, the degree is the 360th part of the circumference.

Descendant.—The western horizon.

Detriment.—A planet located in the sign opposite to his own "house," is said to be in his "detriment." Thus, ♃ when in ♍ is said to be in his detriment.

Diameter, Apparent.—The angle which the diameter of a heavenly body subtends at any time, varying inversely with its distance. The *semi-diameter* of the Sun and Moon, is given daily in the *Nautical Almanac.* In computing the *arc of direction* of a zodiacal parallel, the *semi-diameter* of ☉ or ☽ should always be allowed for, and the arc of direction worked over again, for when falling near the tropic the *arc of duration* of a zodiacal parallel, from the first to the last contact, is very great.

Dignities.—Those situations in which a planet is found to be most powerful.

Direct Motion.—A celestial body is said to have *direct* motion when it advances in the order of the signs of the zodiac, or in the direction of the earth's annual revolution.

Directions.—These are calculations of the arc, or measure of the equator, between any two points in the heavens. See p. 112.

Dispose.—When one planet is found in the dignities of another, the latter is said to dispose of him, thus: ♃ in ♍ is *disposed* of by ☿.

Disposers.—The planets were called the *disposers* by the ancients, as in the first verse of *Genesis* (in the original Hebrew); being regarded as the disposers of all things.

Domal Dignity.—A planet in his own "house," as ☿ in ♊ or ♍.

Dragon's Head and Tail.—The Moon's *north node* was called the *Dragon's head,* and her *south node* the *dragon's tail.* " The Moon's course was early discovered to be of a serpentine form; and, when she was found to rise above the plane of the earth's course about the Sun towards the north, she was feigned to pass the head of the dragon. So, when she crossed the ecliptic into south latitude, she was said to go through the dragon's tail." (Zadkiel). The *dragon's head* was considered fortunate, and the *dragon's tail* unfortunate. It is absurd to think that either of these points can exert any influence.

EARTHY SIGNS.—Taurus, Virgo, and Capricornus.

Eclipse.—An obscuration of a heavenly body owing to the interposition of another, or to its passage through the shadow of a larger body. In astrology the eclipses of the Sun and Moon only have any significance. See p. 24.

Ecliptic.—The great circle of the heavens which the Sun appears to describe in the course of the year, in consequence of the earth's motion round him. The plane of the ecliptic or of the earth's path is the general plane of reference in astronomy. It forms an angle of 23° 27' with the equator, which is consequently the measure of the inclination of the earth's axis to that of her orbit or the ecliptic, and is termed the obliquity of the ecliptic.

Elevated.—The planet that is nearest the meridian is elevated above any other with which it may be in aspect.

Elevation of the Pole.—The latitude of the birth-place.

Elongation.—The angular distance of a heavenly body from the Sun, eastward or westward.

Emersion.—The reappearance of a body after undergoing eclipse, or occultation by the Moon.

Ephemeris.—A daily tabular statement of the geocentric longitudes, latitudes, and declinations of the Sun, Moon, and planets. *Zadkiel's Ephemeris* has been regularly published since the year 1840, and has been much improved since 1878. It is the most useful work of its kind, for astrological purposes. In Urania, for 1880, the author gave a more complete *Ephemeris* (the planets' longitudes and declinations being given to *seconds*), but its publication is suspended for the present, owing to paucity of subscribers.

Equation of Time.—In astrology, this signifies the turning into time of the arc of direction. The best method is that of Ptolemy, viz., the luni-solar, which gives *one year* of life for each *degree* of the arc of direction, and one month for every five minutes of arc.

Equinoctial Signs.—Aries and Libra.

Equinoxes.—The two points of intersection of the Ecliptic and the Equator; so called, because on the Sun's arrival at either point, the night is equal in length to the day, throughout the world. That point (♈ 0°) where the Sun crosses the Equator, going *north*, is termed the *vernal* equinox ; and the opposite point (♎ 0°) is the *autumnal* equinox.

Exaltation.—The most powerful dignity of a planet. See the Table of Essential Dignities, p. 175.

FALL.—A planet's *fall* is in the point opposite to its exaltation.

Familiarities.—The same as configurations. Also when two planets have mutual reception.

Feminine Signs.—Taurus, Cancer, Virgo, Scorpio, Capricornus, and Pisces were termed by Ptolemy *feminine* signs. It is an absurd distinction.

Feral.—Persons born with either the sign *Leo* or the latter half of *Sagittarius* rising, were said to be *feral*, or fierce as a lion.

Fiery Signs.—Aries, Leo, and Sagittarius.

Figure.—The map of the heavens drawn for a given moment is sometimes spoken of as the *figure* of the heavens.

Fixed Signs.—Taurus, Leo, Scorpio, and Aquarius.

Fixed Stars.—All the stars, except the planets, are termed *fixed* stars, owing to their apparent fixity in the sky. They only exert influence when within 5° of the Sun, Moon, ascendant or meridian at birth. Only those of the first magnitude, and clusters, are considered in nativities. A table of eminent fixed stars will be found in the first volume of the " Text-Book of Astrology." The tables of fixed stars given in other astrological works are all faulty and misleading.

Fortunes.—The planets Jupiter and Venus, also the

Sun when he is well placed and in configuration with the benefics.

Fruitful Signs.—Cancer, Scorpio, and Pisces.

Frustration.—A term used in *horary* astrology. When one planet applies to another, and before the aspect can be completed, another intervenes and forms an aspect with the one receiving the application. This is said to destroy or *frustrate* the affair which would otherwise be brought about.

GEMINI.—The third *sign* of the zodiac, which the Sun enters about the 21st of May. It is one of the ancient zodiacal constellations.

Genethliacal.—That which applies to the geniture, or nativity.

Geocentric.—As viewed from the centre of the earth.

Giver of Life.—The " hyleg," or that on which life depends.

HELIOCENTRIC.—As seen from, or having reference to the centre of the Sun.

Hemisphere.—Half the surface of the heavens. The celestial equator divides the heavens into two hemispheres, the northern and the southern. The visible hemisphere is that which is always exposed to view.

Houses.—These are described at pp. 11, 136.

Human Signs.—Gemini, Virgo, and Aquarius. An eclipse falling in one of these signs is said to pre-signify important effects on the human race.

Hyleg.—That point which carries with it the life.

Hylegliacal Places are the first, seventh, ninth, tenth, and eleventh houses of the heavens. See p. 80.

IMMERSION.—The disappearance of a heavenly body when undergoing eclipse.

Impeded or *Impedited.*—A term applied to the Moon when she is in conjunction, square, or opposition with either the Sun, Mars, or Saturn.

M

Imum Cœli.—The lower meridian, or fourth house.

Inconjunct.—When a heavenly body has no familiarity with another.

Increasing in Light.—When the Moon or any planet is leaving the Sun, until the opposition is reached: it is a good testimony.

Inferior Conjunction of Mercury or Venus.—The planet is said to be in inferior conjunction when it is situated in the same longitude as the Sun, and between that luminary and the earth.

Inferior Planets.—Mercury and Venus, which revolve in orbits interior to that of the earth.

Infortunes.—Mars, Saturn, and Uranus. The Moon and Mercury become infortunes when configurated solely with either Mars, Saturn, or Uranus.

Ingresses.—When a planet passes over a point in the zodiac to which the Sun, Moon, Midheaven, or Ascendant has arrived by direction, it is said to ingress thereon. An ingress has not much effect unless it takes place near the birthday-anniversary.

Intercepted.—A sign lying between the cusps of two houses.

LATITUDE.—The distance of any point in the heavens north or south of the ecliptic.

Leo.—The *fifth* sign of the zodiac, which the Sun enters about the 22nd of July; it is one of the ancient zodiacal constellations.

Libra.—The *seventh* sign of the zodiac, which the Sun enters about the 21st of September; it is one of the ancient zodiacal constellations. The commencement of the *sign* Libra, where the equator intersects the ecliptic, is called the autumnal equinox.

Light of Time.—The ⊙ by day, and ☽ by night.

Limb.—The border of the disc of the Sun, Moon, or a planet.

Logarithms.—Artificial numbers of great use in astronomical calculations.

Longitude, Geocentric.—The angular distance of a heavenly body from the first point of *Aries*, measured upon the ecliptic, as viewed from the earth.

Longitude, Heliocentric.—The angular distance of a body from the first point of *Aries*, measured upon the ecliptic, as seen from the Sun.

Longitude of a Place upon the Earth's surface, is the arc intercepted between its meridian and that of some other fixed station used as a line of reference. The English astronomers and geographers reckon their longitudes from the meridian of the Greenwich Royal Observatory ; the whole circumference being divided into 360°, or 24 hours, each hour corresponding to 15°.

Longitude of Perihelion.—The longitude, as viewed from the Sun, of that point of the orbit of a planet which is nearest to him. It is one of the elements of an orbit.

Lord.—That planet which has rule in any sign, as his " house," is called the *lord* of that sign. The lord of the year is that planet ruling the ascending sign at the Solar ingress into *Aries*. The lord of an eclipse is the planet ruling the sign in which it falls.

Luminaries.—The Sun and Moon.

Lunation.—A lunar period, or synodical revolution.

MAGNITUDES OF STARS.—Their relative degree of brightness. The fixed stars are arranged into classes according to their intensity of light. The first six classes include all those which are distinctly visible to the naked eye.

Malefics.—Saturn, Uranus, and Mars. Saturn is the " greater infortune " of the ancients.

Masculine Planets.—The ancients reckoned the Sun, Mars, Jupiter, and Saturn as masculine; Mercury as convertible according to position; Venus and the Moon as feminine.

Masculine Signs.—Aries, Gemini, Leo, Libra, Sagittarius, and Aquarius.

Mututine.—Those stars which rise before the Sun were called matutine.

Mean Noon.—The time when the centre of the *mean* Sun is on the meridian.

Meridian.—The great circle of the heavens passing through the zenith and the poles. The plane of the meridian is the plane of this great circle, and its intersection with the sensible horizon is called the meridian line.

Mundane Aspects are formed by the semi-arcs of the celestial bodies. (See p. 42 of Vol. I of the "Text-Book of Astrology.")

Mundane Parallels are equal distances from the meridian.

NADIR.—The point immediately beneath an observer; it is one of the poles of the rational horizon, the opposite pole being the zenith.

Nebula.—A cluster of stars so closely congregated as to require very powerful telescopes to separate them, and appearing in smaller instruments as cloud-like spots.

Neomenium.—The new Moon.

Nocturnal Arc.—The space through which any celestial body passes while below the horizon.

Nodes.—Those points in the orbit of a planet where it intersects the Ecliptic. The *ascending node* (☊) is the point where it passes from the south to the north side of the Ecliptic; the *descending node* (☋) is the opposite point, where the latitude changes from north to south.

Northern Signs.—Aries, Taurus, Gemini, Cancer, Leo, and Virgo.

Nutation.—An oscillatory motion of the Earth's axis, due chiefly to the action of the Moon upon the spheroidal figure of our globe.

OBLIQUE ASCENSION.—See *Ascension, Oblique.*

Oblique Sphere.—One in which all circles parallel to the Equator are oblique to the horizon, and form acute angles with it.

Obliquity of the Ecliptic.—The inclination of the Ecliptic to the Equator, which amounts at present to about 23° 27'.

Occidental.—Western.

Occultation.—As a general term, implies the eclipsing of one heavenly body by another. It is, however, commonly used to denote the eclipses of stars and planets by the Moon.

Opposition.—A heavenly body is said to be in opposition to another when their longitudes differ 180°, or half the circumference.

Orbit.—The path described by a planet about the Sun.

Orb of a Planet.—The distance within which a planetary aspect continues in force. The orb of the ☉ is 17°, ☽ 12°, ☿ 7°, ♀ 8°, ♂ 7°, ♃ 12°, ♄ 9°, ♅ 5°, ♆ unknown.

Oriental.—Eastern. When a star is to the east of the upper meridian (M.C.), it is said to be oriental ; when to the west, occidental.

PARALLAX.—An apparent change in the position of a celestial body, arising from a change of the observer's station. It diminishes the altitude of an object in the vertical circle. Its effect is greatest in the horizon, where it it is termed the *horizontal* parallax, and disappears altogether in the zenith.

Parallels of Declination.—These are secondary circles, parallel to the celestial equator. In astrology the

parallels of declination—termed, also, zodiacal parallels
—are very potent.

Parents.—In nativities the ☉ and ♄ relate to the
father; the ☽ and ♀ to the mother. The fortune and
length of life of the parents are believed to be pre-
signified by the accidental positions and configurations
of those bodies in the nativity of a child. See pp. 120-
123 of the first vol. of the " Text-Book of Astrology."

Pars Fortunæ.—The " Part of Fortune." Invented
by Ptolemy. It is that point of the horoscope whereon
the rays of the Sun and Moon converge, and where the
Moon would be if the Sun were exactly rising. In
nativities it is an absurdity. In horary questions
relating to money, it is usual to consider its position.
The place of the *pars fortunæ* (⊕) is thus found (in
horary figures):—From the longitude of the ☽ subtract
that of the ☉, and to the difference add the long. of the
ascendant; the sum is the long. of ⊕.

Partile.—An exact or perfect aspect.

Peregrine.—A peregrine planet is one having no
kind of essential dignity. In horary astrology, in a
question relating to stolen goods, a peregrine planet in
the seventh house was taken as the significator of the
thief.

Perigee.—That point in the orbit of a heavenly body
where it is nearest to the earth.

Perihelion.—That point in the orbit of a planet
which is nearest to the Sun.

Periodical Lunation.—The time required by the
Moon to return to her own place in the horoscope, viz.
27^{d} 7^{h} 41^{m}.

Pisces.—The *twelfth* sign of the zodiac, which the
Sun enters about the 21st of February. It is one of
the ancient zodiacal constellations.

Planet, Minor.—The minor planets are small bodies

revolving between the orbits of Mars and Jupiter, which have all been discovered since the beginning of this century. They were termed *asteroids* by Sir W. Herschel.

Planet, Primary.—The primary planets which are found to exert influence are seven in number, viz., Mercury, Venus, Mars, Jupiter, Saturn, Uranus, and Neptune. The asteroids are not considered. Uranus was discovered by Sir W. Herschel, on the 13th of March, 1781. Neptune was discovered on the 23rd of September, 1846, in consequence of the calculations of M. Le Verrier and Mr. Adams, who had found, from the anomalous movements of Uranus, that a distant planet must exist nearly in the position wherein Neptune was situated.

Planetary Hours.—These are described at p. 128.

Planisphere.—The celestial sphere projected on a plane surface.[2]

Platic.—This means *wide*. It is used to denote an aspect within half the sum of the orbs of the two bodies casting the rays which form such aspect.

Pleiades.—A remarkable cluster of stars in the last decanate of *Taurus*, seven or eight of which are visible to the naked eye; the telescope reveals more than two hundred. Job refers to the "*mild* influences of the Pleiades."

Promittor.—That planet which promises to produce the event. The planet applying to the significator, or to which the significator applies. In Nativities the planet to which the ☉, ☽, ☿, M.C., or Asc. applies is the promittor.

---

[2] The best extant is "The Planisphere and Treatise," by H. B. London: J. E. Catty, Paternoster Row.

Proper Motion.—That which is *direct* in the order of the signs of the zodiac.

Prorogator.—The planet which upholds life.

Pythagorean System, now called the *Copernican.*

QUADRATURE, *or Quartile.*—A difference of 90° in the longitudes of two celestial bodies. It is often termed the *square* aspect.

Querent.—The person who inquires, or asks the horary question.

Quesited.—The person or thing inquired about.

Quincunx.—A difference of long. of 150°.

Quindecile.—A difference of long. of 24°.

Quintile.—A difference of long. of 72°.

RADICAL.—That which pertains to the *radix*, or nativity. In horary astrology a figure is *radical* if the same *sign* ascends as at birth. If the ☽ be in the last 3° of a sign, or if the first or last 2° of a sign ascend, the figure was considered unfit for judgment.

Radical Elections.—Times chosen for commencing new undertakings. These will be fully treated of in the second volume of the " Text-Book of Astrology."

Rapt Motion.—The daily apparent motion of the heavens from east to west.

Rapt Parallels.—Equal distances from the meridian formed by rapt motion and measured by the semi-arcs of the bodies directed.

Rays.—Beams of influence constituting *aspects.*

Reception.—The disposing of one planet by another.

Rectification.—The discovery of the true (astrological) moment of birth by comparing the periods of events with the directions that should produce them. See pp. 205-214 of the first volume of the " Text-Book of Astrology."

Refraction.—Owing to the property which the air possesses, a ray of light from a star, in entering the earth's atmosphere, is bent downwards towards its surface, and therefore reaches the eye of an observer with a different direction to that it would have if no atmosphere existed. This *refraction* causes all the celestial bodies to appear at a greater altitude above the horizon than they really are ; and the accurate numerical determination of the amount of refraction is of the greatest importance in many classes of astronomical observations.

Refranation.—When one of two planets, approaching a mutual aspect, falls retrograde before the aspect can be completed. In horary astrology this is held to pre-signify that the event, otherwise promised by the aspect, will come to nothing.

Retrogradation.—An apparent motion of the planets contrary to the order of the signs and to their orbital motion. In nativities it was formerly held that a *retrograde* planet can do little or no good unless otherwise well dignified. It is an absurd notion, and is now abandoned. In *horary* astrology a retrograde planet pre-signifies that nothing promised by it can be relied upon.

Revolutions, Solar.—The return of the Sun to his place at birth. The aspects formed at a Solar revolution, especially those to the radical places of the planets, denote the general influences throughout the ensuing year of life. See p. 120.

Right Ascension.—See *Ascension, Right.*

Right Sphere.—One in which all the circles that are parallel to the equator form right angles with the horizon.

SAGITTARIUS.—The *ninth* sign of the zodiac, which the Sun enters about the 21st of November; it is one of the ancient zodiacal constellations.

Saros.—See *Cycle of Eclipses.*

Satellites.—These are the secondary planets or Moons, which revolve about some of the primary planets; the Moon is a satellite of the earth. Jupiter is accompanied by four satellites, discovered in 1610 by Galileo. Saturn has eight satellites. Mars has two satellites. Uranus has four at least, possibly six. Neptune has but one satellite.

Scheme.—A map of the heavens.

Scorpio.—The *eighth* sign of the zodiac, which the Sun enters about the 22nd of October; it is one of the zodiacal constellations.

Secondary Directions.—Those aspects formed, after birth, to the ☉, ☽, Asc. and M.C. They may be traced in the *Ephemeris* for the year of birth. Every day is reckoned as equal to a year of life. They are much weaker than *primary directions*, and of very slight importance.

Semi-Arc.—Half a diurnal or nocturnal arc.

Semi-quartile.—A difference of 45° in longitude between two bodies.

Semi-sextile.—A difference of 30° in longitude between two bodies.

Separation.—When two planets having been in partile aspect with each other begin to move away therefrom.

Sesqui-quare.—The difference of 135° in longitude between two bodies.

Sextile.—A difference of 60° in longitude.

Significator.—The planet ruling the ascendant is significator of the querent, in horary astrology; also, any planet which happens to be within 5° of the eastern horizon. In nativities, the *significators* are the ☉, ☽, ☿, Asc., and M.C.

Signs of Long Ascension.—Cancer, Leo, Virgo, Libra, Scorpio, and Sagittarius. These are so termed because they take a longer time to ascend than the rest. A sextile aspect falling in these signs was said to have the effect of a square, but this notion is not supported by experience.

Signs of Short Ascension.—Capricornus, Aquarius, Pisces, Aries, Taurus, and Gemini. A trine aspect formed in these signs was said to have the same effect as a square; but this notion is not warranted by experience.

Southern Signs.—Libra, Scorpio, Sagittarius, Capricornus, Aquarius, and Pisces.

Southing.—The meridian transit of a celestial body.

Speculum.—A table comprising the chief data by means of which primary directions are computed in a nativity. It usually contains the latitude, declination, right-ascension, meridian distance (in right-ascension), and semi-arc of each planet.

Sphere.—The figure formed by the rotation of a circle.

Stationary.—When a planet appears to have no motion among the stars it is said to be *stationary*. In nativities, the effect is very powerful when a superior planet is *stationary* in the place of the ☉, ☽, Asc., or M.C., especially when it occurs near the birthday-anniversary.

Succedent.—Those houses which follow the angles, viz. the second, fifth, eighth, and eleventh.

Superior Planets.—Mars, Jupiter, Saturn, Uranus, and Neptune.

Sympathy.—When the significators in the nativity of one person happen to be in the same places in the zodiac as in the nativity of another, there is mutual sympathy between such persons. The strongest is when the ☉ in the one nativity is in the place of the ☽ in the other.

Syzigy.—Either conjunction or opposition, in reference to the orbit of the Moon.

TABLES OF HOUSES.—Tables of houses for London, Edinburgh, Calcutta, and New York will be found in this Appendix. They are computed for the present obliquity of the ecliptic.

Taurus.—The *second* sign of the zodiac, which the Sun enters about the 20th of April; it is one of the ancient zodiacal constellations.

Term.—An essential dignity.

Testimony.—Any aspect or position of a significator in horary questions, bearing on the affair inquired about. In nativities, the positions of the several planets, as regards the ☉, ☽, Asc., and M.C., are testimonies of good or evil, according to their nature.

Time, Apparent, or *Apparent Solar Time,* is the time resulting from an observation of the Sun.

Time, Mean, or *Mean Solar Time.*—"The interval between the times of transit of the Sun over the meridian on successive days is not always the same; and consequently the length of the true solar day varies, the cause of the variation being the unequal progress of the Sun in the ecliptic. But in order to have an equable measure of time astronomers suppose a *mean Sun* to revolve with the real Sun's mean or average motion in the ecliptic; and a clock regulated by this fictitious Sun shows *mean time.* The difference between *apparent* and *mean* time is called the *equation of time,* the clock being sometimes before the Sun, *i.e.* showing *noon* before the true Sun arrives on the meridian, and at others after."—*Hind.*

Time, Sidereal, is the time shown by a clock regulated by the fixed stars. The sidereal day is $3^m 56^s$ shorter than the mean solar day; and hence sidereal time gains upon mean time by this amount daily.

Transits.—In judicial astrology, *transits* are the passing of the planets over the places of the ☉, ☽, Asc. and M.C. at birth. The effects are supposed to vary according to the strength of the planet in transit. Transits near the birthday-anniversary are the most effectual, and it is on these that the force of the Solar Revolution depends. The transits of Saturn endure for some weeks, the second transit (after retrogradation) being usually much more powerful than the first.

Translation of Light.—When one planet separates from the conjunction or aspect of another, and soon after forms the conjunction or aspect with a third, it is said to *translate the light* of the one it leaves to the third. In horary astrology, it denotes aid by a person described by the planet translating the light.

Tredecile.—A difference of longitude of 108°.

Trigons.—These are the four *triplicities*, viz., the fiery, ♈, ♌, ♐ ; the earthy, ♉, ♍, ♑ ; the airy, ♊, ♎, ♒ ; and the watery, ♋, ♏, and ♓.

Trine.—The difference of longitude of 120°.

Triplicity.—See *Trigons.*

Tropical Signs.—Cancer and Capricornus.

ULTRA-ZODIACAL.—Beyond the limits of the Zodiac. A term occasionally applied to the minor planets, some of which pass without the zodiac in the course of their revolution round the Sun.

Under the Sunbeams.—Within 17° of the Sun.

Urania.—A monthly magazine of astrology, meteorology, and physical science was published, in 1880, by the author, under this title. It is at present suspended, but will be republished provided a sufficient number of subscribers put down their names. One of the minor planets is named *Urania.*

VIGINTILE.—The difference of longitude of 18°.

Virgo.—The *sixth* sign of the Zodiac, which the Sun

enters about the 21st of August ; it is one of the ancient zodiacal constellations.

Void of Course.—When the Sun, Moon, or planets form no aspect before leaving a sign.

Zenith.—The point directly over head ; it is the pole of the horizon.

Zenith Distance.—The angular distance of a heavenly body from the zenith.

Zodiac.—A zone or belt of the heavens extending 9° on either side of the *Ecliptic*, and therefore 18° in breadth, within which the Sun and all the larger planets revolve. The Zodiac was divided by the ancients first into *ten* signs, *Libra* being omitted, *Virgo* and *Scorpio* being merged into one. Ptolemy hands down to us the division of the Zodiac into twelve signs, each measuring 30° along the Ecliptic : Aries, Taurus, Gemini, Cancer, Leo. Virgo, Libra, Scorpio, Sagittarius, Capricornus, Aquarius, and Pisces. (See the chapter on the Signs and Constellations of the Zodiac, in the first vol. of the " Text-Book of Astrology.")

Zodiacal Aspects.—Those aspects (angles) measured in degrees of the Zodiac. See *Aspects*.

Zodiacal Parallels.—The parallels of declination.

TABLE OF THE ESSENTIAL DIGNITIES OF THE PLANETS.

Signs.	Houses	Exaltations.		Triplicities. D.N.		Terms of the Planets.										Phases of the Planets.						Detriment.	Fall.
♈	♂ D.	☉	19°	☉	♃	♃	6°	♀	14°	☿	21°	♂	26°	♄	30°	♂	10°	☉	10°	♀	10°	♀	♄
♉	♀ N.	☽	3	♀	☽	♀	8	☿	15	♃	22	♄	26	♂	30	☿	10	☽	10	♄	10	♂	
♊	☿ D.			♄	☿	☿	7	♃	14	♀	21	♄	25	♂	30	♃	10	♂	10	☉	10	♃	
♋	☽ D.N.	♃	15	♂	♂	♂	6	♃	13	☿	20	♀	27	♄	30	♀	10	☿	10	☽	10	♄	♂
♌	☉ D.N.			☉	♃	♄	6	☿	13	♀	19	♃	25	♂	30	♄	10	♃	10	♂	10	♄	
♍	☿ N.	☿	15	♀	☽	☿	7	♀	13	♃	18	♄	24	♂	30	☉	10	♀	10	☿	10	♃	♀
♎	♀ D.	♄	21	♄	☿	♄	6	♀	11	♃	19	☿	24	♂	30	☽	10	♄	10	♃	10	♂	☉
♏	♂ N.			♂	♂	♂	6	♃	14	♀	21	☿	27	♄	30	♂	10	☉	10	♀	10	♀	☽
♐	♃ D.			☉	♃	♃	8	♀	14	☿	19	♄	25	♂	30	☿	10	☽	10	♄	10	☿	
♑	♄ N.	♂	28	♀	☽	♀	6	☿	12	♃	19	♂	25	♄	30	♃	10	♂	10	☉	10	☽	♃
♒	♄ D.			♄	☿	♄	6	☿	12	♀	20	♃	25	♂	30	♀	10	☿	10	☽	10	☉	
♓	♃ N.	♀	27	♂	♂	♀	8	♃	14	☿	20	♂	26	♄	30	♄	10	♃	10	♂	10	☿	☿

TABLES OF HOUSES.

1 Right Ascension of Meridian.	GREENWICH. Lat. 51° 28′ 38″ North.						EDINBURGH. Lat. 55° 57′ 23″ North.					
	Ascendant. ♋	2nd ♌	3rd ♍	4th ♎	5th ♏	6th ♐	Ascendant. ♌	2nd ♌	3rd ♍	4th ♎	5th ♏	6th ♐
H. M. S.	° ′	°	°	°	°	°	° ′	°	°	°	°	°
0 0 0	26 34	12	3	0	9	22	0 30	15	4	0	11	27
0 3 40	27 13	13	3	1	10	23	1 7	16	4	1	12	28
0 7 20	27 54	14	4	2	11	24	1 45	17	5	2	13	29
0 11 1	28 34	15	5	3	12	25	2 22	18	6	3	14	30
0 14 41	29 14	15	6	4	13	25	2 50	19	7	4	15	♑
0 18 21	29 53	16	7	5	14	26	3 36	19	8	5	16	1
0 22 2	0♌33	17	8	6	15	27	4 13	20	8	6	17	2
0 25 42	1 13	18	8	7	16	28	4 49	20	9	7	18	3
0 29 23	1 53	18	9	8	17	29	5 27	21	10	8	19	4
0 33 4	2 33	19	10	9	18	♑	6 4	21	11	9	20	5
0 36 45	3 13	20	11	10	19	1	6 41	22	12	10	21	5
0 40 27	3 53	20	12	11	20	1	7 18	23	13	11	23	6
0 44 8	4 32	21	13	12	22	2	7 55	23	13	12	24	7
0 47 50	5 11	22	14	13	23	3	8 32	24	14	13	25	7
0 51 32	5 51	23	15	14	24	4	9 9	25	15	14	26	8
0 55 14	6 31	23	15	15	25	5	9 46	25	16	15	27	9
0 58 57	7 10	24	16	16	26	6	10 23	26	17	16	28	10
1 2 40	7 49	25	17	17	27	6	11 0	27	17	17	29	11
1 6 24	8 29	26	18	18	28	7	11 37	28	18	18	♐	12
1 10 7	9 8	26	19	19	29	8	12 11	28	19	19	1	12
1 13 51	9 48	27	19	20	♐	9	12 51	29	20	20	2	13
1 17 36	10 28	28	20	21	1	10	13 28	♍	21	21	3	14
1 21 21	11 7	28	21	22	2	10	14 5	0	21	22	4	15
1 25 6	11 47	29	22	23	3	11	14 42	1	22	23	5	15
1 28 52	12 27	♍	23	24	4	12	15 19	2	23	24	6	16
1 32 33	13 7	1	24	25	5	13	15 56	2	24	25	7	17
1 36 25	13 47	1	25	26	6	14	16 33	3	25	26	8	18
1 40 13	14 26	2	25	27	7	14	17 11	4	26	27	9	18
1 44 1	15 6	3	26	28	7	15	17 48	4	26	28	10	19
1 47 49	15 46	4	27	29	8	16	18 25	5	27	29	11	20
1 51 38	16♌27	4	28	30	9	17	19♌ 3	6	28	30	12	21

TABLES OF HOUSES.

Right Ascension of Meridian.	Calcutta. Lat. 22° 33′ 25″ North. Ascendant.		2nd	3rd	4th	5th	6th	New York. Lat. 40° 42′ 42″ North. Ascendant.		2nd	3rd	4th	5th	6th
	♋		♌	♍	♎	♏	♐	♋		♌	♍	♎	♏	♐
H. M. S.	°	′	°	°	°	°	°	°	′	°	°	°	°	°
0 0 0	9	23	4	0	0	4	8	18	54	9	1	0	6	15
0 3 40	10	12	4	0	1	5	9	19	39	9	2	1	7	16
0 7 20	11	1	5	1	2	6	10	20	24	10	3	2	8	17
0 11 1	11	50	6	2	3	7	11	21	8	11	4	3	9	18
0 14 41	12	39	7	3	4	8	12	21	53	12	5	4	10	19
0 18 21	13	28	8	4	5	9	13	22	38	12	6	5	11	20
0 22 2	14	17	8	5	6	10	14	23	22	13	6	6	13	21
0 25 42	15	6	9	6	7	11	15	24	6	14	7	7	14	22
0 29 23	15	54	10	7	8	12	15	24	51	15	8	8	15	22
0 33 4	16	43	11	8	9	13	16	25	36	15	9	9	16	23
0 36 45	17	32	12	9	10	14	17	26	21	16	10	10	17	24
0 40 27	18	20	13	10	11	15	18	27	5	17	11	11	18	25
0 44 8	19	9	14	11	12	16	19	27	50	18	12	12	19	26
0 47 50	19	58	14	12	13	17	20	28	34	19	13	13	20	27
0 51 32	20	47	15	12	14	18	21	29	18	19	13	14	21	28
0 55 14	21	35	16	13	15	19	22	0♌	2	20	14	15	22	28
0 58 57	22	23	17	14	16	20	22	0	46	21	15	16	23	29
1 2 40	23	12	18	15	17	20	23	1	30	22	16	17	24	♑
1 6 24	24	2	19	16	18	21	24	2	15	23	17	18	25	1
1 10 7	24	51	20	17	19	22	25	2	59	23	18	19	26	2
1 13 51	25	40	20	18	20	23	26	3	43	24	19	20	27	3
1 17 36	26	29	21	19	21	24	27	4	27	25	20	21	23	4
1 21 21	27	18	22	20	22	25	28	5	11	26	21	22	29	4
1 25 6	28	7	23	21	23	26	29	5	55	27	22	23	♐	5
1 28 52	28	57	24	22	24	27	29	6	40	27	22	24	1	6
1 32 38	29	46	25	23	25	28	♑	7	21	28	23	25	2	7
1 36 25	0♌	36	26	24	26	29	1	8	9	29	24	26	2	8
1 40 13	1	26	27	25	27	♐	2	8	54	♍	25	27	3	9
1 44 1	2	16	28	26	28	1	3	9	38	1	26	28	4	9
1 47 49	3	6	28	27	29	2	4	10	23	1	27	29	5	10
1 51 38	3♌	56	29	28	30	3	5	11♌	8	2	23	30	6	11

TABLES OF HOUSES.

3 Right Ascension of Meridian.	GREENWICH. Lat. 51° 28′ 38″ North.						EDINBURGH. Lat. 55° 57′ 23″ North.					
	Ascendant.	2nd	3rd	4th	5th	6th	Ascendant.	2nd	3rd	4th	5th	6th
	♌	♍	♍	♏	♐	♑	♌	♍	♍	♏	♐	♑
H. M. S.	° ′	°	°	°	°	°	° ′	°	°	°	°	°
1 51 38	16 27	4	28	0	9	17	19 3	6	28	0	12	21
1 55 27	17 7	5	29	1	10	18	19 40	7	29		13	22
1 59 18	17 47	6	♎	2	11	19	20 18	7	♎	2	14	22
2 3 8	18 27	7	1	3	12	19	20 56	8	1	3	15	23
2 7 0	19 7	8	2	4	13	20	21 34	9	2	4	16	24
2 10 52	19 48	9	2	5	14	21	22 12	10	2	5	17	25
2 14 44	20 29	9	3	6	15	22	22 51	10	3	6	18	25
2 18 37	21 9	10	4	7	16	22	23 29	11	4	7	19	26
2 22 31	21 50	11	5	8	17	23	24 7	12	5	8	20	27
2 26 26	22 31	11	6	9	18	24	24 45	12	6	9	21	28
2 30 21	23 12	12	7	10	19	25	25 23	13	7	10	22	28
2 34 17	23 53	13	8	11	20	25	26 2	14	8	11	22	29
2 38 14	24 35	14	9	12	21	26	26 41	15	8	12	23	♒
2 42 11	25 16	15	10	13	22	27	27 20	16	9	13	24	1
2 46 9	25 58	15	11	14	23	28	27 59	16	10	14	25	1
2 50 9	26 40	16	12	15	24	29	28 39	17	11	15	25	2
2 54 7	27 22	17	12	16	25	29	29 18	18	12	16	27	3
2 58 8	28 4	18	13	17	26	♒	29 57	18	13	17	28	4
3 2 8	28 45	18	14	18	26	1	0♍37	19	14	18	29	4
3 6 10	29 27	19	15	19	27	2	1 16	20	15	19	♑	5
3 10 12	0♍9	20	16	20	28	3	1 56	21	16	20	1	6
3 14 16	0 52	21	17	21	29	3	2 36	21	16	21	2	7
3 18 19	1 35	22	18	22	♑	4	3 16	22	17	22	3	7
3 22 24	2 18	22	19	23	1	5	3 56	23	18	23	4	8
3 26 29	3 1	23	20	24	2	6	4 37	24	19	24	5	9
3 30 35	3 44	24	21	25	3	7	5 17	24	20	25	5	10
3 34 42	4 27	25	22	26	4	7	5 58	25	21	26	6	10
3 38 49	5 10	26	23	27	5	8	6 39	26	22	27	7	11
3 42 57	5 53	27	24	28	6	9	7 19	27	23	28	8	12
3 47 6	6 37	27	25	29	7	10	8 0	28	24	29	9	13
3 51 16	7♍21	28	25	30	8	11	8♍41	28	25	30	10	14

TABLES OF HOUSES.

4 Right Ascension of Meridian.	CALCUTTA. Lat. 22° 33′ 25″ North.						NEW YORK. Lat. 40° 42′ 42″ North.					
	Ascendant.	2nd	3rd	4th	5th	6th	Ascendant.	2nd	3rd	4th	5th	6th
	♌	♌	♍	♏	♐	♑	♌	♍	♍	♏	♐	♑
H. M. S.	° ′	°	°	°	°	°	° ′	°	°	°	°	°
1 51 33	3 56	29	28	0	3	4	11 8	2	28	0	6	11
1 55 27	4 46 ♍	29	1	4	5	11 53	3	29	1	7	12	
1 59 18	5 37	1 ♎	2	5	6	12 38	4	♎	2	8	13	
2 3 8	6 28	2	1	3	6	7	13 23	5	1	3	9	14
2 7 0	7 19	3	2	4	7	8	14 8	5	2	4	10	14
2 10 52	8 10	4	3	5	8	9	14 53	6	3	5	11	15
2 14 44	9 1	4	4	6	9	9	15 39	7	4	6	12	16
2 18 37	9 52	5	5	7	10	10	16 24	8	5	7	13	17
2 22 31	10 43	6	6	8	11	11	17 16	9	5	8	14	18
2 26 26	11 35	7 ♐	7	9	12	12	17 56	10	6	9	15	19
2 30 21	12 27 ♐ 8	8	10	13	13	18 42	10	7	10	16	20	
2 34 17	13 19	9	9	11	13	14	19 28	11	8	11	17	20
2 38 14	14 12	10	10	12	14	15	20 14	12	9	12	18	21
2 42 11	15 4	11	11	13	15	16	21 0	13	10	13	19	22
2 46 9	15 57	12	12	14	16	16	21 47	14	11	14	19	23
2 50 9	16 50	13	13	15	17	17	22 34	15	12	15	20	24
2 54 7	17 43	14	14	16	18	18	23 21	16	13	16	21	25
2 58 8	18 36	15	15	17	19	19	24 8	17	14	17	22	25
3 2 8	19 30	16	16	18	20	20	24 55	17	15	18	23	26
3 6 10	20 24	17	17	19	21	21	25 42	18	16	19	24	27
3 10 12	21 18	18	18	20	22	22	26 29	19	17	20	25	28
3 14 16	22 12	19	19	21	23	23	27 17	20	18	21	26	29
3 18 19	23 6	20	20	22	24	24	28 5	21	19	22	27	≈
3 22 24	24 1	21	21	23	25	25	28 53	22	20	23	28	1
3 26 29	24 56	22	22	24	25	26	29 41	23	21	24	29	1
3 30 35	25 51	23	23	25	26	26	0♍29	21	22	25	♑	2
3 34 42	26 46	24	24	26	27	27	1 17	24	23	26	1	3
3 38 49	27 42	25	25	27	28	28	2 6	25	24	27	2	4
3 42 57	28 33	26	26	23	29	29	2 55	26	25	28	3	5
3 47 6	29 34	27	27	29	♑	≈	3 44	27	26	29	4	6
3 51 16	0♍30	28	28	30	1	1	4♍33	28	27	30	5	7

TABLES OF HOUSES.

5 Right Ascension of Meridian.	GREENWICH. Lat. 51° 28′ 38″ North.						EDINBURGH. Lat. 55° 57′ 23″ North.					
	Ascendant.	2nd	3rd	4th	5th	6th	Ascendant.	2nd	3rd	4th	5th	6th
	♍	♍	♎	♐	♑	♒	♍	♍	♎	♐	♑	♒
H. M. S.	° ′	°	°	°	°	°	° ′	°	°	°	°	°
3 51 16	7 21	28	25	0	8	11	8 41	28	25	0	10	14
3 55 26	8 5	29	26	1	9	12	9 22	29	25	1	11	14
3 59 37	8 49	♎	27	2	10	12	10 3	♎	26	2	12	15
4 3 48	9 33	1	28	3	10	13	10 45	1	27	3	13	16
4 8 1	10 17	2	29	4	11	14	11 27	2	28	4	14	17
4 12 13	11 1	2	♏	5	12	15	12 9	2	29	5	15	17
4 16 27	11 46	3	1	6	13	16	12 51	3	♏	6	16	18
4 20 41	12 30	4	2	7	14	17	13 33	4	1	7	16	19
4 24 55	13 15	5	3	8	15	17	14 15	5	2	8	17	20
4 29 11	14 0	6	4	9	16	18	14 57	6	3	9	18	21
4 33 26	14 45	7	5	10	17	19	15 39	6	4	10	19	21
4 37 22	15 30	8	6	11	18	20	16 21	7	5	11	20	22
4 41 59	16 15	8	7	12	19	21	17 4	8	6	12	21	23
4 46 17	17 0	9	8	13	20	21	17 47	9	6	13	22	24
4 50 34	17 45	10	9	14	21	22	18 30	10	7	14	23	25
4 54 52	18 31	11	10	15	22	23	19 12	10	8	15	24	25
4 59 11	19 16	12	11	16	23	24	19 54	11	9	16	25	26
5 3 30	20 2	13	12	17	24	25	20 37	12	10	17	26	27
5 7 49	20 48	14	13	18	25	26	21 21	13	11	18	27	28
5 12 9	21 34	14	14	19	25	27	22 4	14	12	19	28	29
5 16 29	22 20	15	14	20	26	28	22 47	14	13	20	29	29
5 20 49	23 5	16	15	21	27	28	23 30	15	14	21	29	♓
5 25 10	23 51	17	16	22	28	29	24 13	16	15	22	♒	1
5 29 30	24 37	18	17	23	29	♓	24 56	17	16	23	1	2
5 33 51	25 23	19	18	24	♒	1	25 40	18	17	24	2	3
5 38 13	26 9	20	19	25	1	2	26 23	18	18	25	3	3
5 42 34	26 55	20	20	26	2	3	27 6	19	19	26	4	4
5 46 55	27 41	21	21	27	3	4	27 49	20	19	27	5	5
5 51 17	28 27	22	22	28	4	4	28 33	21	20	28	6	6
5 55 38	29 13	23	23	29	5	5	29 16	22	21	29	7	7
6 0 0	30♍ 0	24	24	30	6	6	30♍ 0	23	22	30	8	7

TABLES OF HOUSES.

G Right Ascension of Meridian.	CALCUTTA. Lat. 22° 33' 25" North.						NEW YORK. Lat. 40° 42' 42" North.					
	Ascendant ♍	2nd ♍	3rd ♎	4th ♐	5th ♑	6th ♒	Ascendant ♍	2nd ♍	3rd ♎	4th ♐	5th ♑	6th ♒
H. M. S.	° '	°	°	°	°	°	° '	°	°	°	°	°
3 51 16	0 36	28	28	0	1	1	4 33	28	27	0	5	7
3 55 26	1 26	29	29	1	2	2	5 22	29	28	1	5	8
3 59 37	2 23	♎	m	2	3	3	6 11	♎	29	2	6	8
4 3 48	3 21	1	1	3	4	4	7 1	1	m	3	7	9
4 8 1	4 18	2	2	4	5	5	7 51	2	1	4	8	10
4 12 13	5 15	3	3	5	6	6	8 41	3	2	5	9	11
4 16 27	6 13	4	5	6	7	7	9 30	4	3	6	10	12
4 20 41	7 11	5	6	7	8	8	10 20	4	4	7	11	13
4 24 55	8 9	6	7	8	9	9	11 10	5	5	8	12	14
4 29 11	9 7	7	8	9	10	10	12 1	6	6	9	13	15
4 33 26	10 5	8	9	10	11	11	12 51	7	7	10	14	16
4 37 22	11 4	9	10	11	12	12	13 41	8	8	11	15	16
4 41 59	12 3	10	11	12	13	13	14 32	9	9	12	16	17
4 46 17	13 2	11	12	13	14	14	15 23	10	10	13	17	18
4 50 34	14 1	12	13	14	15	15	16 14	11	11	14	18	19
4 54 52	15 0	13	14	15	16	16	17 5	12	12	15	19	20
4 59 11	15 59	14	15	16	17	17	17 56	13	13	16	20	21
5 3 30	16 58	15	16	17	18	18	18 47	14	14	17	21	22
5 7 49	17 58	17	17	18	19	19	19 39	15	15	18	22	23
5 12 9	18 58	18	18	19	20	20	20 30	16	16	19	23	24
5 16 29	19 58	19	19	20	21	21	21 22	17	17	20	24	25
5 20 49	20 58	20	20	21	22	22	22 13	18	18	21	25	26
5 25 10	21 58	21	21	22	23	23	23 5	19	19	22	26	27
5 29 30	22 58	22	22	23	24	24	23 57	19	20	23	27	27
5 33 51	23 58	23	23	24	25	25	24 49	20	21	24	28	28
5 38 13	24 58	24	24	25	26	26	25 40	21	22	25	29	29
5 42 34	25 58	25	25	26	27	27	26 32	22	23	26	♒	♓
5 46 55	26 58	26	26	27	28	28	27 24	23	24	27	1	1
5 51 17	27 59	27	27	28	29	29	28 16	24	25	28	1	2
5 55 38	28 59	28	28	29	♒	♓	29 8	25	26	29	2	3
6 0 0	30 ♍ 0	29	29	30	1	1	30 ♍ 0	26	27	30	3	4

TABLES OF HOUSES.

7 Right Ascension of Meridian.	GREENWICH. Lat. 51° 28′ 38″ North.						EDINBURGH. Lat. 55° 57′ 23″ North.					
	Ascendant.	2nd	3rd	4th	5th	6th	Ascendant.	2nd	3rd	4th	5th	6th
	♎	♎	♏	♑	♒	♓	♎	♎	♏	♑	♒	♓
H. M. S.	° ′	°	°	°	°	°	° ′	°	°	°	°	°
6 0 0	0 0	24	24	0	6	6	0 0	23	22	0	8	7
6 4 22	0 47	25	25	1	7	7	0 44	23	23	1	9	8
6 8 43	1 33	26	26	2	8	8	1 27	24	24	2	10	9
6 13 5	2 19	27	27	3	9	9	2 11	25	25	3	11	10
6 17 26	3 5	27	28	4	10	10	2 54	26	26	4	11	11
6 21 47	3 51	28	29	5	11	10	3 37	27	27	5	12	12
6 26 9	4 27	29	♐	6	12	11	4 20	27	28	6	13	12
6 30 30	5 23	♏	1	7	13	12	5 4	28	29	7	14	13
6 34 50	6 9	1	2	8	14	13	5 47	29	♐	8	15	14
6 39 11	6 55	2	3	9	15	14	6 30	♏	1	9	16	15
6 43 31	7 40	2	4	10	16	15	7 13	1	2	10	17	16
6 47 51	8 26	3	4	11	16	16	7 56	1	2	11	18	16
6 52 11	9 12	4	5	12	17	16	8 39	2	3	12	19	17
6 56 30	9 58	5	6	13	18	17	9 23	3	4	13	20	18
7 0 49	10 44	6	7	14	19	18	10 6	4	5	14	21	19
7 5 8	11 29	7	8	15	20	19	10 48	5	6	15	22	20
7 9 26	12 15	8	9	16	21	20	11 30	5	7	16	23	21
7 13 43	13 0	8	10	17	22	21	12 13	6	8	17	24	21
7 18 1	13 45	9	11	18	23	22	12 56	7	9	18	24	22
7 22 18	14 30	10	12	19	24	22	13 39	8	10	19	25	23
7 26 34	15 15	11	13	20	25	23	14 21	9	11	20	26	24
7 30 49	16 0	12	14	21	26	24	15 3	9	12	21	27	25
7 35 5	16 45	13	15	22	27	25	15 45	10	13	22	28	25
7 39 19	17 30	13	16	23	28	26	16 27	11	14	23	29	26
7 43 33	18 14	14	17	24	29	27	17 9	12	14	24	♓	27
7 47 47	18 59	15	18	25	♓	28	17 51	13	15	25	1	28
7 51 59	19 43	16	19	26	1	28	18 33	13	16	26	2	29
7 56 12	20 27	17	20	27	2	29	19 15	14	17	27	3	29
8 0 23	21 11	18	20	28	3	♈	19 57	15	18	28	4	♈
8 4 34	21 55	18	21	29	4	1	20 38	16	19	29	5	1
8 8 44	22♎39	19	22	30	5	2	21♎19	16	20	30	5	2

TABLES OF HOUSES.

8 Right Ascension of Meridian.	CALCUTTA. Lat. 22° 33′ 25″ North.						NEW YORK. Lat. 40° 42′ 42″ North.					
	Ascendant.	2nd	3rd	4th	5th	6th	Ascendant.	2nd	3rd	4th	5th	6th
	♎	♎	♏	♑	♒	♓	♎	♎	♏	♑	♒	♓
H. M. S.	° ′	°	°	°	°	°	° ′	°	°	°	°	°
6 0 0	0 0	29	29	0	1	1	0 0	26	27	0	3	4
6 4 22	1 1	♏	♐	1	2	2	0 52	27	28	1	4	5
6 8 43	2 1	1	1	2	3	3	1 44	28	29	2	5	6
6 13 5	3 2	2	2	3	4	4	2 36	29	29	3	6	7
6 17 26	4 2	3	3	4	5	5	3 28	♏	♐	4	7	8
6 21 47	5 2	4	4	5	6	6	4 20	1	1	5	8	9
6 26 9	6 2	5	5	6	7	7	5 11	2	2	6	9	10
6 30 30	7 2	6	6	7	8	8	6 3	3	3	7	10	11
6 34 50	8 2	7	7	8	9	9	6 55	3	4	8	11	11
6 39 11	9 2	8	8	9	10	10	7 47	4	5	9	12	12
6 43 31	10 2	9	9	10	11	11	8 38	5	6	10	13	13
6 47 51	11 2	10	10	11	12	12	9 30	6	7	11	14	14
6 52 11	12 2	11	11	12	13	14	10 21	7	8	12	15	15
6 56 30	13 2	12	12	13	14	15	11 13	8	9	13	16	16
7 0 49	14 1	13	13	14	15	16	12 4	9	10	14	17	17
7 5 8	15 0	14	14	15	16	17	12 55	10	11	15	18	18
7 9 26	15 59	15	15	16	17	18	13 46	11	12	16	19	19
7 13 43	16 58	16	16	17	18	19	14 37	12	13	17	20	20
7 18 1	17 57	17	17	18	19	20	15 28	13	14	18	21	21
7 22 18	18 56	18	18	19	20	21	16 19	14	15	19	22	22
7 26 34	19 55	19	19	20	21	22	17 9	14	16	20	23	23
7 30 49	20 53	20	20	21	22	23	17 59	15	17	21	24	24
7 35 5	21 50	21	21	22	23	24	18 50	16	18	22	25	25
7 39 19	22 49	22	22	23	24	25	19 40	17	19	23	26	26
7 43 33	23 47	23	23	24	25	26	20 30	18	20	24	27	26
7 47 47	24 45	24	24	25	27	27	21 19	19	21	25	28	27
7 51 59	25 42	25	25	26	28	28	22 9	20	22	26	29	28
7 56 12	26 39	26	26	27	29	29	22 59	21	23	27	X	29
8 0 23	27 37	27	27	28	X	♈	23 49	22	24	28	1	♈
8 4 31	28 34	28	28	29	1	1	24 38	22	25	29	2	1
8 8 44	29 ♎30	29	29	30	2	2	25 ♎27	23	25	30	3	2

TABLES OF HOUSES.

9 Right Ascension of Meridian.	GREENWICH. Lat. 51° 28′ 38″ North.						EDINBURGH. Lat. 55° 57′ 23″ North.					
	Ascendant. ♎	2nd ♏	3rd ♐	4th ♒	5th ♓	6th ♈	Ascendant. ♎	2nd ♏	3rd ♐	4th ♒	5th ♓	6th ♈
H. M. S.	° ′	°	°	°	°	°	° ′	°	°	°	°	°
8 8 44	22 39	19	22	0	5	2	21 19	16	20	0	5	2
8 12 54	23 23	20	23	1	5	3	22 0	17	21	1	6	2
8 17 3	24 7	21	24	2	6	3	22 41	18	22	2	7	3
8 21 11	24 50	22	25	3	7	4	23 21	19	23	3	8	4
8 25 19	25 33	23	26	4	8	5	24 2	20	24	4	9	5
8 29 25	26 16	23	27	5	9	6	24 43	20	25	5	10	6
8 33 31	26 59	24	28	6	10	7	25 23	21	25	6	11	6
8 37 36	27 42	25	29	7	11	8	26 4	22	26	7	12	7
8 41 41	28 25	26	♑	8	12	8	26 44	23	27	8	13	8
8 45 44	29 8	27	1	9	13	9	27 24	23	28	9	14	9
8 49 48	29 51	27	2	10	14	10	28 4	24	29	10	15	9
8 53 50	0♏33	28	3	11	15	11	28 44	25	♑	11	15	10
8 57 52	1 15	29	4	12	16	12	29 23	26	1	12	16	11
9 1 52	1 56	♐	4	13	17	12	0♏3	26	2	13	17	12
9 5 53	2 38	1	5	14	18	13	0 42	27	3	14	18	12
9 9 51	3 20	2	6	15	18	14	1 21	28	4	15	19	13
9 13 51	4 2	2	7	16	19	15	2 1	29	5	16	20	14
9 17 49	4 44	3	8	17	20	16	2 40	29	6	17	21	15
9 21 46	5 25	4	9	18	21	16	3 19	♐	7	18	22	15
9 25 43	6 7	5	10	19	22	17	3 58	1	8	19	22	16
9 29 39	6 48	5	11	20	23	18	4 37	2	9	20	23	17
9 33 34	7 29	6	12	21	24	19	5 15	2	9	21	24	18
9 37 29	8 10	7	13	22	25	19	5 53	3	10	22	25	18
9 41 23	8 51	8	14	23	26	20	6 31	4	11	23	26	19
9 45 16	9 31	9	15	24	27	21	7 9	5	12	24	27	20
9 49 8	10 12	9	16	25	28	22	7 48	5	13	25	28	21
9 53 0	10 53	10	17	26	28	23	8 26	6	14	26	28	21
9 56 52	11 33	11	18	27	29	23	9 4	7	15	27	29	22
10 0 42	12 13	12	19	28	♈	24	9 42	8	16	28	♈	23
10 4 33	12 53	12	20	29	1	25	10 20	8	17	29	1	23
10 8 22	13♏33	13	20	30	2	26	10♏57	9	18	30	2	24

TABLES OF HOUSES.

10 Right Ascension of Meridian	CALCUTTA. Lat. 22° 33' 25" North.						NEW YORK. Lat. 40° 42' 42" North.					
	Ascendant.	2nd	3rd	4th	5th	6th	Ascendant.	2nd	3rd	4th	5th	6th
	♎	♏	♐	♒	♓	♈	♎	♏	♐	♒	♓	♈
H. M. S.	° '	°	°	°	°	°	° '	°	°	°	°	°
8 8 44	29 30	29	29	0	2	2	25 27	23	25	0	3	2
8 12 54	0♏26	♐	♑	1	3	3	26 16	24	26	1	4	3
8 17 3	1 22	1	1	2	4	4	27 5	25	27	2	5	4
8 21 11	2 18	2	2	3	5	5	27 54	26	28	3	6	5
8 25 19	3 14	3	3	4	6	6	28 43	27	29	4	7	6
8 29 25	4 9	4	4	5	7	7	29 31	28	♐	5	8	6
8 33 31	5 4	4	5	6	8	8	0♏19	29	1	6	9	7
8 37 36	5 59	5	5	7	9	9	1 7	29	2	7	10	8
8 41 41	6 54	6	6	8	10	10	1 55	♐	3	8	11	9
8 45 44	7 48	7	7	9	11	11	2 43	1	4	9	12	10
8 49 48	8 42	8	8	10	12	12	3 31	2	5	10	13	11
8 53 50	9 36	9	9	11	13	13	4 18	3	6	11	14	12
8 57 52	10 30	10	10	12	14	14	5 5	4	7	12	15	13
9 1 52	11 24	11	11	13	15	15	5 52	5	8	13	16	13
9 5 53	12 17	12	12	14	16	16	6 39	5	9	14	17	14
9 9 51	13 10	13	13	15	17	17	7 26	6	10	15	18	15
9 13 51	14 3	14	14	16	18	18	8 13	7	11	16	19	16
9 17 49	14 56	14	15	17	19	19	9 0	8	11	17	20	17
9 21 46	15 48	15	16	18	20	20	9 46	9	12	18	21	18
9 25 43	16 41	16	17	19	21	21	10 32	10	13	19	22	19
9 29 39	17 33	17	18	20	22	22	11 18	11	14	20	23	20
9 33 34	18 25	18	19	21	23	23	12 4	11	15	21	24	20
9 37 29	19 17	19	20	22	24	24	12 50	12	16	22	25	21
9 41 23	20 8	20	20	23	25	25	13 36	13	17	23	26	22
9 45 16	20 59	21	21	24	26	26	14 21	14	18	24	26	23
9 49 8	21 50	21	22	25	27	26	15 7	15	19	25	27	24
9 53 0	22 41	22	23	26	28	27	15 52	16	20	26	28	25
9 56 52	23 32	23	24	27	29	28	16 37	16	21	27	29	25
10 0 42	24 23	24	25	28	♈	29	17 22	17	22	28	♈	26
10 4 33	25 14	25	26	29	1	♉	18 7	18	23	29	1	27
10 8 22	26♏ 4	26	27	30	2	1	18♏52	19	24	30	2	28

TABLES OF HOUSES.

11 Right Ascension of Meridian.	GREENWICH. Lat. 51° 28′ 38″ North.						EDINBURGH. Lat. 55° 57′ 23″ North.					
	Ascendant. ♏	2nd ♐	3rd ♑	4th ♓	5th ♈	6th ♈	Ascendant. ♏	2nd ♐	3rd ♑	4th ♓	5th ♈	6th ♈
H. M. S.	° ′	°	°	°	°	°	° ′	°	°	°	°	°
10 8 22	13 33	13	20	0	2	26	10 57	9	18	0	2	24
10 12 11	14 14	14	21	1	3	26	11 35	10	19	1	3	25
10 15 59	14 54	15	22	2	4	27	12 12	11	20	2	4	26
10 19 47	15 34	16	23	3	5	28	12 49	12	21	3	4	26
10 23 35	16 13	16	24	4	5	29	13 27	12	22	4	5	27
10 27 22	16 53	17	25	5	6	29	14 4	13	23	5	6	28
10 31 8	17 33	18	26	6	7	♉	14 41	14	24	6	7	28
10 34 54	18 13	19	27	7	8	1	15 18	14	25	7	8	29
10 38 39	18 53	20	23	8	9	2	15 55	15	26	8	9	30
10 42 24	19 32	20	29	9	10	2	16 32	16	27	9	9	♉
10 46 9	20 12	21	≈	10	11	3	17 9	17	28	10	10	1
10 49 53	20 52	22	1	11	11	4	17 46	18	29	11	11	2
10 53 36	21 31	23	2	12	12	4	18 23	18	≈	12	12	3
10 57 20	22 11	24	3	13	13	5	19 0	19	1	13	13	3
11 1 3	22 50	24	4	14	14	6	19 37	20	2	14	13	4
11 4 46	23 29	25	5	15	15	7	20 14	21	3	15	14	5
11 8 28	24 9	26	6	16	16	7	20 51	22	4	16	15	5
11 12 10	24 49	27	8	17	17	8	21 28	22	5	17	16	6
11 15 52	25 28	28	9	18	17	9	22 5	23	6	18	17	7
11 19 33	26 7	29	10	19	18	10	22 42	24	7	19	18	7
11 23 15	26 47	30	11	20	19	10	23 19	25	8	20	18	8
11 26 56	27 27	♑	12	21	20	11	23 56	25	9	21	19	9
11 30 37	28 7	1	13	22	21	12	24 33	26	10	22	20	9
11 34 18	28 47	2	14	23	22	13	25 11	27	12	23	21	10
11 37 58	29 27	3	15	24	23	13	25 47	28	13	24	22	11
11 41 39	0 ♐ 7	4	16	25	23	14	26 24	29	14	25	22	11
11 45 19	0 46	5	17	26	24	15	27 1	30	15	26	23	12
11 48 59	1 26	5	18	27	25	15	27 38	♑	16	27	24	13
11 52 40	2 6	6	19	28	26	16	28 5	1	17	28	25	14
11 56 20	2 46	7	20	29	26	17	28 53	2	18	29	26	14
12 0 0	3 ♐ 26	8	21	30	27	17	29 ♏ 30	3	19	30	26	15

TABLES OF HOUSES.

12 Right Ascension of Meridian	CALCUTTA. Lat. 22° 33′ 25″ North.						NEW YORK. Lat. 40° 42′ 42″ North.					
	Ascendant.	2nd	3rd	4th	5th	6th	Ascendant.	2nd	3rd	4th	5th	6th
	♏	♐	♑	♓	♈	♉	♏	♐	♑	♓	♈	♈
H. M. S.	° ′	°	°	°	°	°	° ′	°	°	°	°	°
10 8 22	26 4	26	27	0	2	1	18 52	19	24	0	2	28
10 12 11	26 54	27	28	1	3	2	19 37	20	25	1	3	29
10 15 59	27 44	28	29	2	4	3	20 22	21	26	2	4	29
10 19 47	28 34	28	♒	3	5	4	21 6	21	27	3	5	♉
10 23 35	29 24	29	1	4	6	5	21 51	22	28	4	6	1
10 27 22	0 ♐ 14	♑	2	5	7	6	22 36	23	29	5	7	2
10 31 8	1 3	1	3	6	8	6	23 20	24	♒	6	8	3
10 34 54	1 53	2	4	7	9	7	24 5	25	1	7	8	4
10 38 39	2 42	3	5	8	10	8	24 49	26	1	8	9	4
10 42 24	3 31	4	5	9	11	9	25 33	26	2	9	10	5
10 46 9	4 20	4	6	10	12	10	26 17	27	3	10	11	6
10 49 53	5 9	5	7	11	13	11	27 1	28	4	11	12	7
10 53 36	5 58	6	8	12	14	12	27 45	29	5	12	13	7
10 57 20	6 48	7	9	13	15	13	28 30	♑	6	13	14	8
11 1 3	7 37	8	10	14	16	13	29 14	1	7	14	15	9
11 4 46	8 25	9	11	15	17	14	29 58	2	8	15	16	10
11 8 28	9 13	10	12	16	18	15	0 ♐ 42	2	9	16	17	11
11 12 10	10 2	11	13	17	19	16	1 26	3	10	17	17	11
11 15 52	10 51	11	14	18	19	17	2 10	4	11	18	18	12
11 19 33	11 40	12	15	19	20	18	2 55	5	12	19	19	13
11 23 15	12 28	13	16	20	21	18	3 39	6	13	20	20	14
11 26 56	13 17	14	17	21	22	19	4 24	7	14	21	21	15
11 30 37	14 6	15	18	22	23	20	5 9	8	15	22	22	15
11 34 18	14 54	16	19	23	24	21	5 54	9	16	23	23	16
11 37 58	15 43	17	20	24	25	22	6 38	9	17	24	24	17
11 41 39	16 32	18	21	25	26	23	7 22	10	18	25	24	18
11 45 19	17 21	19	22	26	27	24	8 7	11	19	26	25	18
11 48 59	18 10	19	23	27	28	24	8 52	12	21	27	26	19
11 52 40	18 59	20	24	28	29	25	9 36	13	22	28	27	20
11 56 20	19 48	21	25	29	30	26	10 21	14	23	29	28	21
12 0 0	20 ♐ 37	22	26	30	30	27	11 ♐ 6	15	24	30	29	21

TABLES OF HOUSES.

13	GREENWICH. Lat. 51° 28′ 38″ North.						EDINBURGH. Lat. 55° 57′ 23″ North.					
Right Ascension of Meridian.	Ascendant.	2nd	3rd	4th	5th	6th	Ascendant.	2nd	3rd	4th	5th	6th
	♐	♑	♒	♈	♈	♉	♏	♑	♒	♈	♈	♉
H. M. S.	° ′	°	°	°	°	°	° ′	°	°	°	°	°
12 0 0	3 26	8	21	0	27	17	29 30	3	19	0	26	15
12 3 40	4 7	9	22	1	28	18	0 ♐ 7	4	20	1	27	16
12 7 20	4 47	10	24	2	29	19	0 44	5	22	2	28	16
12 11 1	5 28	11	25	3	♉	20	1 22	6	23	3	29	17
12 14 41	6 9	12	26	4	1	20	1 59	7	24	4	♉	18
12 18 21	6 49	13	27	5	1	21	2 37	7	25	5	1	18
12 22 2	7 30	14	28	6	2	22	3 15	8	26	6	2	19
12 25 42	8 12	15	29	7	3	23	3 53	9	27	7	3	20
12 29 23	8 54	16	♓	8	4	23	4 31	10	29	8	4	20
12 33 4	9 36	17	2	9	5	24	5 10	11	♓	9	4	21
12 36 45	10 18	18	3	10	6	25	5 51	12	1	10	5	22
12 40 27	11 0	19	4	11	6	25	6 30	13	3	11	6	22
12 44 8	11 43	20	5	12	7	26	7 9	14	4	12	7	23
12 47 50	12 25	21	6	13	8	27	7 48	15	5	13	7	24
12 51 32	13 8	22	7	14	9	28	8 28	16	6	14	8	24
12 55 14	13 52	23	9	15	10	28	9 8	17	7	15	9	25
12 58 57	14 35	24	10	16	11	29	9 48	18	8	16	9	26
13 2 40	15 19	25	11	17	11	♊	10 28	19	10	17	10	27
13 6 24	16 3	26	12	18	12	1	11 9	20	11	18	11	27
13 10 7	16 47	27	13	19	13	1	11 50	21	12	19	12	28
13 13 51	17 32	28	15	20	14	2	12 31	22	13	20	12	29
13 17 36	18 18	29	16	21	15	3	13 13	23	15	21	13	29
13 21 21	19 4	♒	17	22	16	4	13 55	25	16	22	14	♊
13 25 6	19 50	1	18	23	16	4	14 37	26	17	23	15	1
13 28 52	20 37	2	20	24	17	5	15 20	27	19	24	16	2
13 32 38	21 24	4	21	25	18	6	16 3	28	20	25	16	2
13 36 25	22 11	5	22	26	19	7	16 47	29	22	26	17	3
13 40 13	22 59	6	23	27	20	7	17 31	♒	23	27	18	4
13 44 1	23 48	7	25	28	21	8	18 15	1	24	28	19	4
13 47 49	24 37	8	26	29	21	9	19 0	3	25	29	20	5
13 51 33	25 ♐ 26	10	27	30	22	10	19 ♐ 45	4	27	30	20	6

TABLES OF HOUSES.

14 Right Ascension of Meridian.	CALCUTTA. Lat. 22° 33' 25" North.						NEW YORK. Lat. 40° 42' 42" North.					
	Ascendant. ♐	2nd ♑	3rd ♒	4th ♈	5th ♉	6th ♉	Ascendant. ♐	2nd ♑	3rd ♒	4th ♈	5th ♈	6th ♉
H. M. S.	° '	°	°	°	°	°	° '	°	°	°	°	°
12 0 0	29 37	22	26	0	0	27	11 6	15	24	0	29	21
12 3 40	.1 26	23	27	1	1	27	11 51	16	25	1	0	22
12 7 20	22 15	21	28	2	2	28	12 36	17	26	2	1	23
12 11 1	23 5	24	29	3	3	29	13 21	18	27	3	1	24
12 14 41	23 54	25	♓	4	4	♉	14 7	18	28	4	2	25
12 18 21	24 44	26	1	5	5	1	14 53	19	29	5	3	25
12 22 2	25 33	27	2	6	6	1	15 39	20	♓	6	4	26
12 25 42	26 24	28	3	7	7	2	16 26	21	1	7	5	27
12 29 23	27 14	29	4	8	8	3	17 13	22	2	8	6	28
12 33 4	28 4	♒	5	9	9	4	18 0	23	3	9	7	23
12 36 45	28 54	1	6	10	9	5	18 46	24	4	10	7	29
12 40 27	29 45	2	7	11	10	6	19 33	25	6	11	8	11
12 44 8	0♑36	3	8	12	11	6	20 21	26	7	12	9	1
12 47 50	1 27	4	9	13	12	7	21 9	27	8	13	10	2
12 51 32	2 18	5	10	14	13	8	21 57	28	9	14	11	2
12 55 14	3 10	6	11	15	14	9	22 45	29	16	15	12	3
12 58 57	4 2	7	12	16	15	10	23 33	♒	11	16	13	4
13 2 40	4 54	8	13	17	16	11	24 22	1	12	17	13	5
13 6 24	5 46	9	14	18	17	11	25 12	2	13	18	14	6
13 10 7	6 38	10	15	19	18	12	26 2	3	15	19	15	7
13 13 51	7 31	11	17	20	18	13	26 52	4	16	20	16	7
13 17 36	8 24	12	18	21	19	14	27 43	6	17	21	17	8
13 21 21	9 18	13	19	22	20	15	28 34	7	18	22	18	9
13 25 6	10 12	14	20	23	21	16	29 25	8	19	23	19	10
13 28 52	11 6	15	21	24	22	16	0♑16	9	20	24	19	11
13 32 38	12 0	16	22	25	23	17	1 8	10	21	25	20	11
13 36 25	12 55	17	23	26	24	18	2 1	11	23	26	21	12
13 40 13	13 50	18	24	27	25	19	2 54	12	24	27	22	13
13 44 1	14 46	19	25	28	26	20	3 48	13	25	28	23	14
13 47 49	15 42	20	26	29	26	21	4 42	15	26	29	24	14
13 51 38	16♑38	21	27	30	27	22	5♑37	16	27	30	25	15

TABLES OF HOUSES.

15 Right Ascension of Meridian.	GREENWICH. Lat. 51° 28′ 38″ North.						EDINBURGH. Lat. 55° 57′ 23″ North.					
	Ascendant ♐	2nd ♒	3rd ♓	4th ♉	5th ♉	6th ♊	Ascendant ♐	2nd ♒	3rd ♓	4th ♉	5th ♉	6th ♊
H. M. S.	° ′	°	°	°	°	°	° ′	°	°	°	°	°
13 51 33	25 26	10	27	0	22	10	19 45	4	27	0	20	6
13 55 27	26 16	11	28	1	23	11	20 31	6	28	1	21	7
13 59 18	27 7	12	♈	2	24	11	21 18	7	♈	2	22	7
14 3 8	27 58	14	1	3	25	12	22 5	8	1	3	23	8
14 7 0	28 50	15	2	4	26	13	22 52	10	2	4	24	9
14 10 52	29 43	16	4	5	26	14	23 40	11	4	5	25	10
14 14 44	0♑37	18	5	6	27	15	24 29	13	5	6	25	10
14 18 37	1 31	19	6	7	28	15	25 19	14	7	7	26	11
14 22 31	2 26	20	8	8	29	16	26 9	16	8	8	27	12
14 26 26	3 22	22	9	9	11	17	27 0	17	9	9	28	13
14 30 21	4 19	23	10	10	1	18	27 52	19	11	10	29	14
14 34 17	5 16	25	11	11	2	19	28 45	20	12	11	30	14
14 38 14	6 14	26	13	12	2	20	29 38	22	13	12	11	15
14 42 11	7 13	28	14	13	3	20	0♑32	24	15	13	1	16
14 46 9	8 14	29	15	14	4	21	1 27	25	16	14	2	17
14 50 9	9 16	♓	17	15	5	22	2 25	27	18	15	3	18
14 54 7	10 19	3	18	16	6	23	3 23	29	19	16	4	18
14 58 8	11 23	4	19	17	7	24	4 22	♓	20	17	5	19
15 2 8	12 28	6	21	18	8	25	5 21	2	22	18	5	20
15 6 10	13 34	8	22	19	9	26	6 22	4	23	19	6	21
15 10 12	14 41	9	23	20	9	27	7 25	6	24	20	7	22
15 14 16	15 50	11	24	21	10	27	8 29	8	26	21	8	23
15 18 19	17 0	13	26	22	11	28	9 34	10	27	22	9	24
15 22 24	18 12	14	27	23	12	29	10 41	12	29	23	10	24
15 26 29	19 27	16	28	24	13	♋	11 50	14	♉	24	11	25
15 30 35	20 43	17	29	25	14	1	13 0	16	1	25	12	26
15 34 42	22 0	19	♉	26	15	2	14 12	18	3	26	13	27
15 38 49	23 18	21	2	27	16	3	15 27	20	4	27	13	28
15 42 57	24 38	22	3	28	17	4	16 43	22	5	28	14	29
15 47 6	26 1	24	5	29	18	5	18 2	24	7	29	15	♋
15 51 16	27♑26	26	6	30	18	6	19♑23	26	8	30	16	1

TABLES OF HOUSES.

16 Right Ascension of Meridian.	CALCUTTA. Lat. 22° 33' 25" North.						NEW YORK. Lat. 40° 42' 42" North.					
	Ascendant. ♑	2nd ♒	3rd ♓	4th ♉	5th ♉	6th ♊	Ascendant. ♑	2nd ♒	3rd ♓	4th ♉	5th ♉	6th ♊
H. M. S.	° '	°	°	°	°	°	° '	°	°	°	°	°
13 51 38	16 38	22	27	0	27	22	5 37	16	27	0	25	15
13 55 27	17 35	23	29	1	28	23	6 3.	17	29	1	25	16
13 59 18	18 32	24	♈	2	29	24	7 28	18	♈	2	26	17
14 3 8	19 30	25	1	3	♊	25	8 25	19	1	3	27	18
14 7 0	20 28	26	2	4	1	25	9 22	21	2	4	28	19
14 10 52	21 27	27	3	5	2	26	10 20	22	3	5	29	19
14 14 44	22 26	28	4	6	3	27	11 18	23	5	6	♊	20
14 18 37	23 26	29	5	7	4	28	12 18	24	6	7	1	21
14 22 31	24 26	♓	7	8	4	29	13 18	26	7	8	2	22
14 26 26	25 26	2	8	9	5	♋	14 18	27	8	9	2	23
14 30 21	26 27	3	9	10	6	1	15 19	28	9	10	3	24
14 34 17	27 29	4	10	11	7	2	16 21	♓	11	11	4	25
14 38 14	28 31	5	11	12	8	3	17 25	1	12	12	5	25
14 42 11	29 34	6	12	13	9	4	18 29	2	13	13	6	26
14 46 9	0♒37	7	13	14	10	4	19 34	4	14	14	7	27
14 50 9	1 41	9	14	15	11	5	20 39	5	16	15	8	28.
14 54 7	2 45	10	16	16	12	6	21 45	6	17	16	9	29
14 58 8	3 50	11	17	17	13	7	22 53	8	18	17	10	♋
15 2 8	4 56	12	18	18	14	8	24 2	9	19	18	10	1
15 6 10	6 2	13	19	19	15	9	25 12	11	20	19	11	2
15 10 12	7 9	14	20	20	15	10	26 22	12	22	20	12	3
15 14 16	8 17	16	21	21	16	11	27 33	13	23	21	13	4
15 18 19	9 25	17	23	22	17	12	28 45	15	24	22	14	5
15 22 24	10 34	18	24	23	18	13	29 59	16	25	23	15	6
15 26 29	11 44	20	25	24	19	14	1♒15	18	26	24	16	7
15 30 35	12 54	21	26	25	20	15	2 33	19	28	25	17	8
15 34 42	14 5	22	27	26	21	16	3 55	21	29	26	18	9
15 38 49	15 17	23	28	27	22	17	5 7	22	♉	27	19	10
15 42 57	16 29	25	29	28	23	18	6 28	24	1	28	20	11
15 47 6	17 42	26	♉	29	24	19	7 49	25	3	29	21	12
15 51 16	18♒55	27	2	30	25	20	9♒10	27	4	30	22	13

TABLES OF HOUSES.

17 Right Ascension of Meridian.	GREENWICH. Lat. 51° 28′ 38″ North.						EDINBURGH. Lat. 55° 57′ 23″ North.					
	Ascendant ♑	2nd ♓	3rd ♉	4th ♊	5th ♊	6th ♋	Ascendant ♑	2nd ♓	3rd ♉	4th ♊	5th ♊	6th ♋
H. M. S.	° ′	°	°	°	°	°	° ′	°	°	°	°	°
15 51 16	27 26	26	6	0	18	6	19 23	26	8	0	16	1
15 55 26	28 53	28	7	1	19	7	20 45	28	9	1	17	2
15 59 37	0♒22	♈	9	2	20	8	22 10	♈	11	2	18	3
16 3 48	1 53	1	10	3	21	9	23 40	2	12	3	19	4
16 8 1	3 27	3	11	4	22	10	25 12	4	13	4	20	5
16 12 13	5 3	5	12	5	23	11	26 47	6	14	5	21	6
16 16 27	6 42	7	14	6	24	12	28 25	8	16	6	22	7
16 20 41	8 23	9	15	7	25	13	0♒ 6	10	17	7	23	8
16 24 55	10 7	11	16	8	26	14	1 51	12	18	8	24	9
16 29 11	11 54	12	17	9	27	16	3 40	11	19	9	24	10
16 33 26	13 44	14	18	10	28	17	5 33	16	21	10	25	11
16 37 42	15 37	16	20	11	29	18	7 31	18	22	11	26	12
16 41 59	17 32	18	21	12	♋	19	9 33	20	23	12	27	14
16 46 17	19 30	20	22	13	1	20	11 40	22	24	13	28	15
16 50 34	21 32	21	23	14	2	21	13 51	24	26	14	29	16
16 54 52	23 38	23	25	15	3	22	16 7	26	27	15	♋	17
16 59 11	25 46	25	26	16	4	24	18 28	28	28	16	1	18
17 3 30	27 58	27	27	17	5	25	20 55	♉	29	17	2	19
17 7 49	0♓11	28	28	18	6	26	23 28	2	II	18	3	21
17 12 9	2 27	♉	29	19	7	27	26 5	4	2	19	4	22
17 16 29	4 47	2	II	20	8	29	28 49	6	3	20	5	23
17 20 49	7 9	3	1	21	9	♌	1♓38	8	4	21	6	24
17 25 10	9 35	5	2	22	10	1	4 31	9	5	22	7	26
17 29 30	12 2	7	3	23	11	3	7 28	11	6	23	8	27
17 33 51	14 32	8	5	24	12	4	10 33	13	7	24	9	28
17 38 13	17 3	10	6	25	13	5	13 41	15	9	25	11	♌
17 42 34	19 35	11	7	26	14	7	16 52	16	10	26	12	1
17 46 55	22 9	13	8	27	15	8	20 6	18	11	27	13	3
17 51 17	24 47	14	9	28	16	10	23 23	20	12	28	14	4
17 55 38	27 23	16	10	29	17	11	26 41	21	13	29	15	6
18 0 0	30♓ 0	17	11	30	18	13	30♓ 0	23	14	30	16	7

TABLES OF HOUSES.

18 Right Ascension of Meridian.	CALCUTTA. Lat. 22° 33' 25" North.						NEW YORK. Lat. 40° 42' 42" North.					
	Ascendant. ♒	2nd ♓	3rd ♉	4th II	5th II	6th ♋	Ascendant. ♒	2nd ♓	3rd ♉	4th ♊	5th ♊	6th ♋
H. M. S.	° '	°	°	°	°	°	° '	°	°	°	°	°
15 51 16	18 55	27	2	0	25	20	9 10	27	4	0	22	13
15 55 26	20 10	29	3	1	26	21	10 33	28	5	1	22	14
15 59 37	21 25	Υ	4	2	27	22	11 58	Υ	6	2	23	15
16 3 48	22 40	1	5	3	28	23	13 25	1	7	3	24	16
16 8 1	23 56	2	6	4	29	24	14 53	3	9	4	25	17
16 12 13	25 13	4	7	5	♋	25	16 22	4	10	5	26	18
16 16 27	26 31	5	9	6	1	26	17 53	6	11	6	27	19
16 20 41	27 49	6	10	7	2	27	19 25	8	12	7	28	20
16 24 55	29 8	8	11	8	2	28	20 58	9	13	8	29	21
16 29 11	0 ♓ 27	9	12	9	3	29	22 32	11	15	9	♋	22
16 33 26	1 47	10	13	10	4	♌	24 7	12	16	10	1	23
16 37 42	3 8	12	14	11	5	2	25 45	14	17	11	2	25
16 41 59	4 29	13	15	12	6	3	27 25	15	18	12	3	26
16 46 17	5 50	14	16	13	7	4	29 6	17	19	13	4	27
16 50 34	7 13	16	18	14	8	5	0 ♓ 47	18	20	14	5	28
16 54 52	8 36	17	19	15	9	6	2 29	20	22	15	6	29
16 59 11	10 0	18	20	16	10	7	4 12	21	23	16	7	♌
17 3 30	11 23	19	21	17	11	8	5 57	23	24	17	8	2
17 7 49	12 47	21	22	18	12	10	7 44	24	25	18	9	3
17 12 9	14 12	22	23	19	13	11	9 31	26	26	19	10	4
17 16 29	15 37	23	24	20	15	12	11 19	27	27	20	11	5
17 20 49	17 2	25	25	21	16	13	13 8	29	28	21	12	7
17 25 10	18 27	26	26	22	17	14	14 59	♉	II	22	13	8
17 29 30	19 53	27	27	23	18	15	16 50	2	1	23	14	9
17 33 51	21 20	28	29	24	19	17	18 41	4	2	24	15	10
17 38 13	22 46	♉	II	25	20	18	20 33	5	3	25	16	12
17 42 34	24 13	1	1	26	21	19	22 26	6	4	26	17	13
17 46 55	25 39	2	2	27	22	20	24 19	7	5	27	18	14
17 51 17	27 6	4	3	28	23	21	26 13	9	6	28	20	16
17 55 38	28 33	5	4	29	24	23	28 6	10	7	29	21	17
18 0 0	30 ♓ 0	6	5	30	25	24	30 ♓ 0	12	8	30	22	18

o

19 Right Ascension of Meridian.	GREENWICH. Lat. 51° 28' 38" North.						EDINBURGH. Lat. 55° 57' 23" North.					
	Ascendant.	2nd	3rd	4th	5th	6th	Ascendant.	2nd	3rd	4th	5th	6th
	♈	♉	♊	♋	♋	♌	♈	♉	♊	♋	♋	♌
H. M. S.	° ′	°	°	°	°	°	° ′	°	°	°	°	°
18 0 0	0 0	17	11	0	18	13	0 0	23	14	0	16	7
18 4 22	2 37	19	13	1	20	14	3 19	24	15	1	17	9
18 8 43	5 13	20	14	2	21	16	6 37	26	16	2	18	10
18 13 5	7 41	22	15	3	22	17	9 54	27	17	3	19	12
18 17 26	10 25	23	16	4	23	19	13 8	29	18	4	20	14
18 21 47	12 57	25	17	5	24	20	16 19	♊	19	5	21	16
18 26 9	15 28	26	18	6	25	22	19 27	1	21	6	22	17
18 30 30	17 58	28	19	7	26	23	22 32	3	22	7	24	19
18 34 50	20 25	29	20	8	27	25	25 29	4	23	8	25	21
18 39 11	22 51	♊	21	9	29	27	28 22	6	24	9	26	22
18 43 31	25 13	1	22	10	♌	28	1♉11	7	25	10	27	24
18 47 51	27 33	2	23	11	1	♍	3 55	8	26	11	28	26
18 52 11	29 49	4	24	12	2	2	6 32	9	27	12	♌	28
18 56 30	2♉ 2	5	25	13	3	3	9 5	11	28	13	1	♍
19 0 49	4 14	6	26	14	4	5	11 32	12	29	14	2	2
19 5 8	6 22	8	27	15	6	7	13 53	13	♋	15	3	4
19 9 26	8 28	9	28	16	7	9	16 9	14	1	16	4	6
19 13 43	10 30	10	29	17	8	10	18 20	15	2	17	6	8
19 18 1	12 28	11	♋	18	9	12	20 27	16	3	18	7	10
19 22 18	14 23	12	1	19	10	14	22 29	18	4	19	8	12
19 26 34	16 16	13	2	20	12	16	24 27	19	5	20	9	14
19 30 49	18 6	14	3	21	13	18	26 20	20	6	21	11	16
19 35 5	19 53	16	4	22	14	19	28 9	21	7	22	12	18
19 39 19	21 37	17	5	23	15	21	29 54	22	8	23	13	20
19 43 33	23 18	18	6	24	16	23	1♊35	23	9	24	14	22
19 47 47	24 57	19	7	25	18	25	3 13	24	9	25	15	24
19 51 59	26 33	20	8	26	19	27	4 48	25	10	26	17	26
19 56 12	28 7	21	9	27	20	28	6 20	26	11	27	18	28
20 0 23	29 38	22	10	28	21	♎	7 50	27	12	28	20	♎
20 4 34	1♊ 7	23	11	29	23	2	9 15	28	13	29	21	2
20 8 44	2 34	24	12	30	24	4	10♊37	29	14	30	22	4

TABLES OF HOUSES.

20 Right Ascension of Meridian.	CALCUTTA. Lat. 22° 33' 25" North.						NEW YORK. Lat. 40° 42' 42" North.					
	Ascendant.	2nd	3rd	4th	5th	6th	Ascendant.	2nd	3rd	4th	5th	6th
	♈	♉	♊	♋	♋	♌	♈	♉	♊	♋	♋	♌
H. M. S.	° '	°	°	°	°	°	° '	°	°	°	°	°
18 0 0	0 0	6	5	0	25	24	0 0	12	8	0	22	18
18 4 22	1 27	7	6	1	26	25	1 54	13	9	1	23	20
18 8 43	2 54	9	7	2	27	26	3 47	14	10	2	24	21
18 13 5	4 21	10	8	3	28	28	5 41	16	12	3	25	23
18 17 26	5 47	11	9	4	29	29	7 31	17	13	4	26	24
18 21 47	7 14	12	10	5	♌	♍	9 27	18	14	5	27	25
18 26 9	8 40	13	11	6	1	2	11 19	20	15	6	28	27
18 30 30	10 7	15	12	7	2	3	13 10	21	16	7	29	28
18 34 50	11 33	16	13	8	3	4	15 1	22	17	8	♌	♍
18 39 11	12 58	17	14	9	4	5	16 52	23	18	9	2	1
18 43 31	14 23	18	16	10	5	7	18 41	25	19	10	3	3
18 47 51	15 48	19	17	11	7	8	20 29	26	20	11	4	4
18 52 11	17 13	20	18	12	8	9	22 16	27	21	12	5	6
18 56 30	18 37	22	19	13	9	10	24 3	28	22	13	6	7
19 0 49	20 0	23	20	14	10	12	25 48	11	23	14	7	9
19 5 8	21 21	24	21	15	11	13	27 31	1	24	15	8	10
19 9 26	22 47	25	22	16	12	14	29 13	2	25	16	10	12
19 13 43	24 10	26	23	17	14	15	0♉54	3	26	17	11	13
19 18 1	25 31	27	24	18	15	17	2 35	4	27	18	12	15
19 22 18	26 52	28	25	19	16	18	4 15	6	28	19	13	16
19 26 34	28 13	29	26	20	17	19	5 53	7	29	20	14	18
19 30 49	29 33	♊	27	21	18	20	7 28	8	♋	21	15	19
19 35 5	0♉52	2	28	22	19	22	9 2	9	1	22	17	21
19 39 19	2 11	3	28	23	20	23	10 35	10	2	23	18	22
19 43 33	3 29	4	29	24	21	24	12 7	11	3	24	19	24
19 47 47	4 47	5	♋	25	22	26	13 38	12	4	25	20	25
19 51 59	6 4	6	1	26	24	28	15 7	13	5	26	21	27
19 56 12	7 20	7	2	27	25	29	16 35	14	6	27	23	29
20 0 23	8 35	8	3	28	26	♎	18 2	15	7	28	24	♎
20 4 34	9 50	9	4	29	27	1	19 27	16	8	29	25	2
20 8 44	11♉5	10	5	30	28	3	20♉50	17	9	30	26	3

TABLES OF HOUSES.

21 Right Ascension of Meridian.	GREENWICH. Lat. 51° 28′ 38″ North.						EDINBURGH. Lat. 55° 57′ 23″ North.					
	Ascendant. ♊	2nd ♊	3rd ♋	4th ♌	5th ♌	6th ♎	Ascendant. ♊	2nd ♊	3rd ♋	4th ♌	5th ♌	6th ♎
H. M. S.	° ′	°	°	°	°	°	° ′	°	°	°	°	°
20 8 44	2 34	24	12	0	24	4	10 37	29	14	0	22	4
20 12 54	3 59	25	12	1	25	6	11 58	♋	15	1	23	6
20 17 3	5 22	26	13	2	27	7	13 17	1	16	2	25	8
20 21 11	6 42	27	14	3	28	9	14 33	2	17	3	26	10
20 25 18	8 0	28	15	4	29	11	15 48	3	17	4	27	12
20 29 25	9 17	29	16	5	♍	13	17 0	4	18	5	29	14
20 33 31	10 33	♋	17	6	2	14	18 10	5	19	6	♍	16
20 37 36	11 48	1	18	7	3	16	19 19	6	20	7	1	18
20 41 41	13 0	2	19	8	4	18	20 26	6	21	8	3	20
20 45 44	14 10	3	20	9	6	19	21 31	7	22	9	4	22
20 49 48	15 19	3	21	10	7	21	22 35	8	23	10	5	24
20 53 50	16 26	4	21	11	8	23	23 38	9	24	11	7	26
20 57 52	17 32	5	22	12	9	24	24 39	10	25	12	8	28
21 1 52	18 37	6	23	13	11	26	25 38	11	25	13	10	♏
21 5 53	19 41	7	24	14	12	28	26 37	12	26	14	11	1
21 9 51	20 44	8	25	15	13	29	27 35	12	27	15	12	3
21 13 51	21 46	9	26	16	15	♏	28 33	13	28	16	14	5
21 17 49	22 47	10	27	17	16	2	29 28	14	29	17	15	6
21 21 46	23 46	10	28	18	17	4	0♋22	15	30	18	17	8
21 25 43	24 44	11	28	19	19	5	1 15	16	Ω	19	18	10
21 29 39	25 41	12	29	20	20	7	2 8	16	1	20	19	11
21 33 34	26 38	13	Ω	21	22	8	3 0	17	2	21	21	13
21 37 29	27 34	14	1	22	23	10	3 51	18	3	22	22	14
21 41 23	28 29	15	2	23	24	11	4 41	19	4	23	23	16
21 45 16	29 23	15	3	24	25	13	5 31	20	5	24	25	17
21 49 8	0♋17	16	4	25	26	14	6 20	20	5	25	26	19
21 53 0	1 10	17	4	26	28	15	7 8	21	6	26	28	20
21 56 52	2 2	18	5	27	29	16	7 55	22	7	27	29	22
22 0 42	2 53	19	6	28	♎	18	8 42	23	8	28	♎	23
22 4 33	3 44	19	7	29	2	19	9 29	23	9	29	2	24
22 8 22	4♋34	20	8	30	3	20	10♋15	24	10	30	3	26

TABLES OF HOUSES.

22 Right Ascension of Meridian.	CALCUTTA. Lat. 22° 33′ 25″ North.						NEW YORK. Lat. 40° 42′ 42″ North.					
	Ascendant.	2nd	3rd	4th	5th	6th	Ascendant.	2nd	3rd	4th	5th	6th
	♉	♊	♋	♌	♌	♎	♉	♊	♋	♌	♌	♎
H. M. S.	° ′	°	°	°	°	°	° ′	°	°	°	°	°
20 8 44	11 5	10	5	0	28	3	20 50	17	9	0	26	3
20 12 54	12 18	11	6	1	29	4	22 11	18	9	1	27	5
20 17 3	13 31	12	7	2	♍	5	23 32	19	10	2	29	6
20 21 11	14 43	13	8	3	2	7	24 53	20	11	3	♍	8
20 25 18	15 55	14	9	4	3	8	26 5	21	12	4	1	9
20 29 25	17 6	15	10	5	4	9	27 27	22	13	5	2	11
20 33 31	18 16	16	11	6	5	10	28 45	23	14	6	3	12
20 37 36	19 26	17	12	7	6	12	0♊ 1	24	15	7	5	14
20 41 41	20 35	18	13	8	7	13	1 15	25	16	8	6	15
20 45 44	21 43	19	14	9	9	14	2 26	26	17	9	7	17
20 49 48	22 51	20	15	10	10	15	3 38	27	18	10	8	18
20 53 50	23 58	21	16	11	11	17	4 48	28	19	11	10	19
20 57 52	25 4	22	16	12	12	18	5 58	29	20	12	11	21
21 1 52	26 10	23	17	13	13	19	7 7	♋	20	13	12	22
21 5 53	27 15	24	18	14	14	20	8 15	1	21	14	13	24
21 9 51	28 19	25	19	15	16	21	9 21	2	22	15	14	25
21 13 51	29 23	26	20	16	17	23	10 26	3	23	16	16	26
21 17 49	0♊26	26	21	17	18	24	11 31	4	24	17	17	28
21 21 46	1 29	27	22	18	19	26	12 35	5	25	18	18	29
21 25 43	2 31	28	23	19	20	26	13 39	5	26	19	19	♏
21 29 39	3 33	29	24	20	21	27	14 41	6	27	20	21	2
21 33 34	4 31	♋	25	21	22	28	15 42	7	28	21	22	3
21 37 29	5 34	1	26	22	23	♏	16 42	8	28	22	23	4
21 41 23	6 34	2	26	23	25	1	17 42	9	29	23	24	6
21 45 16	7 34	3	27	24	26	2	18 42	10	♌	24	25	7
21 49 8	8 33	4	28	25	27	3	19 40	11	1	25	27	8
21 53 0	9 32	5	29	26	28	4	20 38	11	2	26	28	9
21 56 52	10 30	5	♌	27	29	5	21 35	12	3	27	29	11
22 0 42	11 28	6	1	28	♎	6	22 32	13	4	28	♎	12
22 4 33	12 25	7	2	29	1	7	23 28	14	5	29	1	13
22 8 22	13♊22	8	3	30	2	8	24♊23	15	5	30	3	14

TABLES OF HOUSES.

23 Right Ascension of Meridian.	GREENWICH. Lat. 51° 28′ 38″ North.						EDINBURGH. Lat. 55° 57′ 23″ North.						
	Ascendant ♋	2nd ♋	3rd ♌	4th ♍	5th ♎	6th ♏	Ascendant ♋	2nd ♋	3rd ♌	4th ♍	5th ♎	6th ♏
H. M. S.	° ′	°	°	°	°	°	° ′	°	°	°	°	°
22 8 22	4 34	20	8	0	3	20	10 15	24	10	0	3	26
22 12 11	5 23	21	8	1	4	21	11 0	25	10	1	4	27
22 15 59	6 12	22	9	2	6	23	11 45	26	11	2	6	28
22 19 47	7 1	23	10	3	7	24	12 29	26	12	3	7	29
22 23 35	7 49	23	11	4	8	25	13 13	27	13	4	8	♐
22 27 22	8 36	24	12	5	9	26	13 57	28	14	5	10	2
22 31 8	9 23	25	13	6	10	28	14 40	28	14	6	11	3
22 34 54	10 10	26	14	7	12	29	15 23	29	15	7	12	4
22 33 39	10 56	26	14	8	13	♐	16 5	♌	16	8	14	5
22 42 24	11 42	27	15	9	14	1	16 47	1	17	9	15	7
22 46 9	12 28	28	16	10	15	2	17 29	1	18	10	16	8
22 49 53	13 13	29	17	11	17	3	18 10	2	18	11	18	9
22 53 36	13 57	29	18	12	18	4	18 51	3	19	12	19	10
22 57 20	14 41	♌	19	13	19	5	19 32	4	20	13	20	11
23 1 3	15 25	1	19	14	20	6	20 12	4	21	14	21	12
23 4 46	16 8	2	20	15	21	7	20 52	5	22	15	23	13
23 8 28	16 52	2	21	16	23	8	21 32	6	22	16	24	14
23 12 10	17 35	3	22	17	24	9	22 12	6	23	17	25	15
23 15 52	18 17	4	23	18	25	10	22 51	7	24	18	26	16
23 19 33	19 0	5	24	19	26	11	23 30	8	25	19	28	17
23 23 15	19 42	5	24	20	27	12	24 9	8	26	20	29	18
23 26 56	20 24	6	25	21	29	13	24 50	9	26	21	♏	19
23 30 37	21 6	7	26	22	♏	14	25 29	10	27	22	1	20
23 34 18	21 48	7	27	23	1	15	26 7	10	28	23	2	21
23 37 58	22 30	8	28	24	2	16	26 45	11	29	24	4	22
23 41 39	23 11	9	28	25	3	17	27 23	12	30	25	5	23
23 45 19	23 51	9	29	26	4	18	28 1	12	♍	26	6	23
23 48 59	24 32	10	♍	27	5	19	28 38	13	1	27	7	24
23 52 40	25 13	11	1	28	6	20	29 16	14	2	28	8	25
23 56 20	25 53	12	2	29	8	21	29 53	14	3	29	10	26
24 0 0	26 34	12	3	30	9	22	0 ♌ 30	15	4	30	11	27

TABLES OF HOUSES.

24 Right Ascension of Meridian.	CALCUTTA. Lat. 22° 33′ 25″ North.						NEW YORK. Lat. 40° 42′ 42″ North.					
H. M. S.	Ascendant ♊	2nd ♋	3rd ♌	4th ♍	5th ♎	6th ♏	Ascendant ♊	2nd ♋	3rd ♌	4th ♍	5th ♎	6th ♏
22 8 22	13 22	8	3	0	2	8	24 23	15	5	0	3	14
22 12 11	14 18	9	4	1	4	10	25 16	16	6	1	4	15
22 15 59	15 14	10	4	2	5	11	26 12	16	7	2	5	17
22 19 47	16 10	11	5	3	6	12	27 6	17	8	3	6	18
22 23 35	17 5	11	6	4	7	13	27 59	18	9	4	7	19
22 27 22	18 0	12	7	5	8	14	28 52	19	10	5	8	20
22 31 8	18 54	13	8	6	9	15	29 44	20	11	6	10	21
22 34 54	19 48	14	9	7	10	16	0♋35	21	11	7	11	22
22 38 39	20 42	15	10	8	11	17	1 26	21	12	8	12	23
22 42 24	21 36	16	11	9	12	18	2 17	22	13	9	13	24
22 46 9	22 29	17	12	10	13	19	3 8	23	14	10	14	25
22 49 53	23 22	17	13	11	14	20	3 58	24	15	11	15	27
22 53 36	24 14	18	13	12	16	21	4 48	25	16	12	17	28
22 57 20	25 6	19	14	13	17	22	5 38	25	17	13	18	29
23 1 3	25 58	20	15	14	18	23	6 27	26	17	14	19	♐
23 4 46	26 50	21	16	15	19	24	7 15	27	18	15	20	1
23 8 28	27 42	22	17	16	20	25	8 3	28	19	16	21	2
23 12 10	28 33	22	18	17	21	26	8 51	29	20	17	22	3
23 15 52	29 24	23	19	18	22	27	9 39	29	21	18	23	4
23 19 33	0♋15	24	20	19	23	28	10 27	♌	22	19	24	5
23 23 15	1 6	25	21	20	24	29	11 14	1	23	20	26	6
23 26 56	1 56	26	21	21	25	♐	12 0	2	23	21	27	7
23 30 37	2 46	26	22	22	26	1	12 47	2	24	22	28	8
23 34 13	3 36	27	23	23	27	1	13 34	3	25	23	29	9
23 37 58	4 27	28	24	24	28	2	14 21	4	26	24	♏	10
23 41 39	5 16	29	25	25	29	3	15 7	5	27	25	1	11
23 45 19	6 6	♌	26	26	♏	4	15 53	5	28	26	2	12
23 48 59	6 55	1	27	27	1	5	16 39	6	29	27	3	12
23 52 40	7 45	2	28	28	2	6	17 24	7	30	28	4	13
23 56 20	8 34	3	29	29	3	7	18 9	8	♍	29	5	14
24 0 0	9♋23	4	30	30	4	8	18♋54	9	1	30	6	15

INDEX.

ADVERTISEMENT.

Cloth lettered, 288 pp., demy 8vo., Price 10s. 6d.,

THE
TEXT-BOOK of ASTROLOGY.

Vol. I.—GENETHLIALOGY.

By ALFRED J. PEARCE,

Author of "THE WEATHER GUIDE-BOOK," ETC.

LONDON: COUSINS AND CO., 3, YORK STREET,
COVENT GARDEN.

CONTENTS.

The "*Athenæum*," *May 3rd, 1879, says:*—"The 'Text-Book' is not at all a catchpenny tract; it is seriously written, and may be perused with advantage by anyone interested in astrology, provided he will maintain a cool judgment. The author, while contending for the dignity of his science, warns his readers against illiterate adventurers who pretend to tell fortunes by its means, and his work shows that its practice requires some degree of education and of labour, for its methods are founded on astronomical calculations.... Mr. Pearce deals with conspicuous examples—kings, queens, and emperors, Prince Albert, the Prince of Wales, the Princess Louise—and this we take to be a proper mode of dealing with a scientific subject in the present conditions of society."

www.ingramcontent.com/pod-product-compliance
Lightning Source LLC
Chambersburg PA
CBHW021706210326
41599CB00013B/1536